D1071022

THE KLUWER INTERNATIONAL SERIES
IN ENGINEERING AND COMPUTER SCIENCE

COMMUNICATIONS AND INFORMATION THEORY

Consulting Editor

Robert G. Gallager

AN INTRODUCTION TO CRYPTOLOGY

by

Henk C.A. van Tilborg
Eindhoven University of Technology
THE NETHERLANDS

KLUWER ACADEMIC PUBLISHERS
Boston / Dordrecht / Lancaster

Distributors for North America
Kluwer Academic Publishers
101 Philip Drive
Assinippi Park
Norwell, Massachusetts 02061 USA

Distributors for the UK and Ireland
Kluwer Academic Publishers
Falcon House, Queen Square
Lancaster LAI IRN, UNITED KINGDOM

Distributors for all other countries
Kluwer Academic Publishers Group
Distribution Centre
Post Office Box 322
3300 AH Dordrecht, THE NETHERLANDS

Library of Congress Cataloging-in-Publication Data

Tilborg, Henk C. A. van, 1947–
  An introduction to cryptology.

  (The Kluwer international series in engineering and
computer science)
  Bibliography: p.
  Includes index.
  1. Cryptology. 2. Cyrptography—Data processing.
I. Title. II. Series.
Z103.T54 1987   652'.8   88–616
ISBN 0–89838–271–8

Copyright © 1988 by Kluwer Academic Publishers

All rights reserved. No part of this publication may be reproduced, stored in a retrieval system or transmitted in any form or by any means, mechanical, photocopying, recording, or otherwise, without the prior written permission of the publisher, Kluwer Academic Publishers, 101 Philip Drive, Assinippi Park, Norwell, Massachusetts 02061.

Printed in the United States of America

to Marijke, who after
so many years still is
an enigma to me

CONTENTS

		page
Contents		vii
Preface		ix
1.	INTRODUCTION	1
2.	CLASSICAL SYSTEMS	7
2.1.	Caesar, simple substitutions, Vigenère	7
2.2.	The incidence of coincidences	11
2.3.	Vernam, Playfair, Transpositions, Hagelin, Enigma	14
3.	SHIFT REGISTER SEQUENCES	19
3.1.	Introduction	19
3.2.	Linear feedback shift registers	22
3.3.	Non-linear algorithms	29
4.	SHANNON THEORY	39
5.	HUFFMAN CODES	47
6.	DES	55
7.	PUBLIC KEY CRYPTOGRAPHY	63

8.   THE DISCRETE LOGARITHM PROBLEM   67
8.1.   The discrete logarithm system   67
8.2.   How to take discrete logarithms   69

9.   RSA   77
9.1.   The RSA system   77
9.2.   The Solovay and Strassen primality test   82
9.3.   The Cohen and Lenstra primality test   84
9.4.   The Rabin variant   89

10.   THE MCELIECE SYSTEM   93

11.   THE KNAPSACK PROBLEM   97
11.1.   The knapsack system   97
11.2.   The Shamir attack   101
11.3.   The Lagarias and Odlyzko attack   105

12.   THRESHOLD SCHEMES   111

13.   OTHER DIRECTIONS   115

Appendix A. ELEMENTARY NUMBER THEORY   119
A.1.   Introduction   119
A.2.   Euclid's Algorithm   121
A.3.   Congruences, Fermat, Euler, Chinese Remainder Theorem   124
A.4.   Quadratic residues   129
A.5.   Möbius inversion formula, the principle of inclusion
        and exclusion   132

Appendix B. THE THEORY OF FINITE FIELDS   137
B.1.   Groups, rings, ideals and fields   137
B.2.   Constructions   142
B.3.   The number of irreducible polynomials over $\mathbb{F}_q$ 144
B.4.   The structure of finite fields   148

References   159
Notations   165
Index   167

PREFACE

Since the appearance in 1976 of W. Diffie and M.E. Hellman's paper "New directions in cryptography" (Dif76) cryptology has gained an enormous popularity in the academic world. The reasons are twofold. The first reason has directly to do with the paper. The two authors showed that it is possible (at least in theory) to exchange data over a public channel in a secure way, i.e. without having to worry that eavesdroppers understand the contents of the messages. Very soon proposals followed for systems that realized these so-called Public Key Cryptosystems. These proposals lead to very interesting mathematical problems of a very wide nature. Areas like number theory, abstract algebra, complexity theory, combinatorial and probabilistic algorithms, information theory and discrete mathematics all play a rôle in a very natural way. As a side bonus students enjoy taking a course in cryptology because on top of the already existing appeal that this discipline has, they all of a sudden see various abstract theories elegantly applied in one single course. The second reason that cryptology attracts so much attention is society's growing need for the protection that cryptosystems can give. Computer controlled communication networks are playing an ever increasing role in our daily life. Their use raises problems regarding privacy and authentication, that need to be solved.

This manuscript consists of three parts. In the first six chapters conventional cryptosystems and related topics are discussed. Various good books on this classical material already exist and have been used by me.

If it were only for this part, I would not have written this manuscript. On the other hand conventional cryptosystems are so widely used, that their theory has to be included in any textbook.

The second seven chapters discuss the new area of public key cryptography. Most existing books discuss this area too briefly. The existing material is scattered over the literature. So the need to bring that knowledge together certainly existed. Some of this material involves areas in mathematics that do not belong to the standard knowledge of most of the interested people. For this group two appendices are included. One on elementary number theory and one on the theory of finite fields. The latter appendix is also heavily used in the chapter on shift register sequences.

At the end of most of the chapters problems have been added. The material in the other chapters is not very suited to problems at a class room level.

This book is based on a course for students in Mathematics and Computing Science, although also some interested Electrical Engineering students, majoring in Information Theory, have taken the course. I would like to thank all of these students for the enthusiasm they have demonstrated for this field and for tracking the numerous misprints, that existed in the first version of this book. I am very much in debt to Jan Willem Nienhuys, Cor van Pul, Jos Remijn and Herman Tiersma. At many places they have improved the manuscript with useful comments. Finally I want to thank Anita Klooster for the patience with which she has typed and retyped this manuscript over and over again.

Henk van Tilborg

AN INTRODUCTION TO CRYPTOLOGY

1 INTRODUCTION

Cryptology, the study of cryptosystems, can be divided into two disciplines. *Cryptography* concerns itself with the design of cryptosystems, while *cryptanalysis* studies the breaking of cryptosystems. At this moment we will not give a formal definition of a cryptosystem. That will come later in this chapter. We assume that the reader has the right intuitive idea of what a cryptosystem is.

Why would anybody use a cryptosystem? There are two possible reasons:

> *Privacy:* When transmitting or storing data, one does not want an eavesdropper to understand the contents of the transmitted or stored messages.

> *Authentication:* This property is the equivalent of a signature. The receiver of a message wants proof that a message comes from a certain party and not from somebody else (even if the original party later wants to deny it).

Throughout the centuries (see [Kah67]) cryptosystems have been used by the military and by the diplomatic services. The nowadays widespread use of computer controlled communication systems in industry or by civil services often asks for special protection of the data by cryptosystems.

Since storing data can be viewed as transmission of the data in the time domain, we shall always use the term transmission when discussing a situation where data are stored and/or transmitted. Before we can describe the *conventional cryptosystem* as formulated by C.E. Shannon [Shn49] we need a more formal description of concepts like languages and texts.

Let A be a finite set, which we will call *alphabet*. $|A|$ will denote the cardinality of A. We shall often use $\mathbb{Z}_q = \{0,1,...,q-1\}$ as alphabet, where we work with its elements modulo q (see the beginning of § A.3 and § B.2). $\mathbb{Z}_{26}$ can be identified with $\{a,b,...,z\}$. In modern day practical situations q

will often be 2 or a power of 2. A concatenation of n letters from $\mathbb{Z}_q$ will be called an *n-gram* and denoted by $\underline{a} = (a_0, a_1, \cdots, a_{n-1})$. The set of all n-grams from $\mathbb{Z}_q$ will be denoted by $\mathbb{Z}_q^n$. A *text* of letters from $\mathbb{Z}_q$ is an element from $\mathbb{Z}_q^* := \bigcup_{n \geq 0} \mathbb{Z}_q^n$. A *language* is a subset of $\mathbb{Z}_q^*$. In the case of programming languages this subset is precisely defined by recursion rules, while in the case of spoken languages these rules are very loose. Here we shall choose a probabilistic approach.

A finite or infinite *plaintext source* S of text called *plaintext* from $\mathbb{Z}_q$ is a finite resp. an infinite sequence of random variables

$$(M_0, M_1, \ldots, M_{r-1})$$

resp.

$$(M_0, M_1, M_2, \cdots)$$

described by the probabilities that events occur. So

$$Pr_{plain}\{M_j = m_0, M_{j+1} = m_1, \ldots, M_{j+n-1} = m_{n-1}\}, \quad j \geq 0,$$

is given for all possible events $(m_0, m_1, \ldots, m_{n-1}) \in \mathbb{Z}_q^n$. In the case that $j=0$, we shall simply write $Pr_{plain}\{(m_0, m_1, \ldots, m_{n-1})\}$. Of course these probabilities must satisfy some obvious relations:

i) $Pr_{plain}\{(m_0, m_1, \ldots, m_{n-1})\} \geq 0$ for all $(m_0, m_1, \ldots, m_{n-1})$.

ii) $\displaystyle\sum_{(m_0, m_1, \ldots, m_{n-1})} Pr_{plain}\{(m_0, m_1, \ldots, m_{n-1})\} = 1$

iii)   *Kolmogorov's consistency condition* :

$$\sum_{(m_n, m_{n+1}, \ldots, m_{l-1})} Pr_{plain}\{(m_0, m_1, \ldots, m_{l-1})\} = Pr_{plain}\{(m_0, m_1, \ldots, m_{n-1})\}, l > n,$$

Example 1.1: S generates 1-grams with an independent but identical distribution, say $p(t), 0 \leq t < q$. So

$$Pr_{plain}\{(m_0, m_1, \cdots, m_{n-1})\} = p(m_0) p(m_1) \cdots p(m_{n-1}), \quad n \geq 1.$$

The distribution of the letters in English texts is given in Table 1.1 (see [Mey82], Table 12-1). In this model one has

$$Pr_{plain}\{(run)\} = Pr_{plain}\{(urn)\} = p(r) p(u) p(n) = 0.0612 \times 0.0271 \times 0.0709 \approx 1.18 * 10^{-4}.$$

a	0.0804	h	0.0549	o	0.0760	u	0.0271
b	0.0154	i	0.0726	p	0.0200	v	0.0099
c	0.0306	j	0.0016	q	0.0011	w	0.0192
d	0.0399	k	0.0067	r	0.0612	x	0.0019
e	0.1251	l	0.0414	s	0.0654	y	0.0173
f	0.0230	m	0.0253	t	0.0925	z	0.0009
g	0.0196	n	0.0709				

Table 1.1 Probability distribution of 1-grams in English language

s\t	a	b	c	d	e	f	g	h	i	j	k	l	m
a	0.0011	0.0193	0.0388	0.0469	0.0020	0.0100	0.0233	0.0020	0.0480	0.0020	0.0103	0.1052	0.0281
b	0.0931	0.0057	0.0016	0.0008	0.3219	0.0000	0.0000	0.0000	0.0605	0.0057	0.0000	0.1242	0.0049
c	0.1202	0.0000	0.0196	0.0004	0.1707	0.0000	0.0000	0.1277	0.0761	0.0000	0.0324	0.0369	0.0015
d	0.1044	0.0020	0.0026	0.0218	0.3778	0.0007	0.0132	0.0007	0.1803	0.0033	0.0000	0.0125	0.0178
e	0.0660	0.0036	0.0433	0.1194	0.0438	0.0142	0.0125	0.0021	0.0158	0.0005	0.0036	0.0456	0.0340
f	0.0838	0.0000	0.0000	0.0000	0.1283	0.0924	0.0000	0.0000	0.1608	0.0000	0.0000	0.0299	0.0009
g	0.1078	0.0000	0.0000	0.0018	0.2394	0.0000	0.0177	0.1281	0.0839	0.0000	0.0000	0.0203	0.0027
h	0.1769	0.0005	0.0014	0.0008	0.5623	0.0000	0.0000	0.0005	0.1167	0.0000	0.0000	0.0016	0.0016
i	0.0380	0.0082	0.0767	0.0459	0.0437	0.0129	0.0280	0.0002	0.0016	0.0000	0.0050	0.0567	0.0297
j	0.1259	0.0000	0.0000	0.0000	0.1818	0.0000	0.0000	0.0000	0.0350	0.0000	0.0000	0.0000	0.0000
k	0.0395	0.0028	0.0000	0.0028	0.5282	0.0028	0.0000	0.0198	0.1582	0.0000	0.0113	0.0198	0.0028
l	0.1342	0.0019	0.0022	0.0736	0.1918	0.0105	0.0108	0.0000	0.1521	0.0000	0.0079	0.1413	0.0082
m	0.1822	0.0337	0.0026	0.0000	0.2975	0.0010	0.0000	0.0000	0.1345	0.0000	0.0010	0.0010	0.0654
n	0.0550	0.0004	0.0621	0.1681	0.1212	0.0102	0.1391	0.0013	0.0665	0.0009	0.0066	0.0073	0.0104
o	0.0085	0.0101	0.0162	0.0231	0.0037	0.1299	0.0082	0.0025	0.0092	0.0014	0.0078	0.0416	0.0706
p	0.1359	0.0000	0.0006	0.0000	0.1747	0.0000	0.0000	0.0237	0.0423	0.0000	0.0000	0.0812	0.0073
q	0.0000	0.0000	0.0000	0.0000	0.0000	0.0000	0.0000	0.0000	0.0000	0.0000	0.0000	0.0000	0.0000
r	0.1026	0.0033	0.0172	0.0282	0.2795	0.0031	0.0175	0.0017	0.1181	0.0000	0.0205	0.0164	0.0303
s	0.0604	0.0012	0.0284	0.0027	0.1795	0.0024	0.0000	0.0561	0.1177	0.0000	0.0091	0.0145	0.0112
t	0.0619	0.0003	0.0036	0.0002	0.1417	0.0007	0.0002	0.3512	0.1406	0.0000	0.0000	0.0101	0.0044
u	0.0344	0.0415	0.0491	0.0243	0.0434	0.0052	0.0382	0.0010	0.0258	0.0000	0.0014	0.1097	0.0329
v	0.0749	0.0000	0.0000	0.0023	0.6014	0.0000	0.0000	0.0000	0.2569	0.0000	0.0000	0.0000	0.0012
w	0.2291	0.0008	0.0000	0.0032	0.1942	0.0000	0.0000	0.1422	0.2104	0.0000	0.0000	0.0041	0.0000
x	0.0672	0.0000	0.1119	0.0000	0.1269	0.0000	0.0000	0.0075	0.1119	0.0000	0.0000	0.0000	0.0075
y	0.0586	0.0034	0.0103	0.0069	0.2897	0.0000	0.0000	0.0000	0.0690	0.0000	0.0034	0.0172	0.0379
z	0.2278	0.0000	0.0000	0.0000	0.4557	0.0000	0.0000	0.0000	0.2152	0.0000	0.0000	0.0127	0.0000

s\t	n	o	p	q	r	s	t	u	v	w	x	y	z
a	0.1878	0.0008	0.0222	0.0000	0.1180	0.1001	0.1574	0.0137	0.0212	0.0057	0.0026	0.0312	0.0023
b	0.0000	0.0964	0.0000	0.0000	0.0662	0.0229	0.0049	0.0727	0.0016	0.0000	0.0000	0.1168	0.0000
c	0.0011	0.2283	0.0000	0.0004	0.0426	0.0087	0.0893	0.0347	0.0000	0.0000	0.0000	0.0094	0.0000
d	0.0053	0.0733	0.0000	0.0007	0.0324	0.0495	0.0013	0.0601	0.0099	0.0040	0.0000	0.0264	0.0000
e	0.1381	0.0040	0.0192	0.0034	0.1927	0.1231	0.0404	0.0048	0.0215	0.0205	0.0152	0.0121	0.0004
f	0.0009	0.2789	0.0000	0.0000	0.1215	0.0026	0.0496	0.0462	0.0000	0.0000	0.0000	0.0043	0.0000
g	0.0451	0.1140	0.0000	0.0000	0.1325	0.0256	0.0247	0.0512	0.0000	0.0000	0.0000	0.0053	0.0000
h	0.0038	0.0786	0.0000	0.0000	0.0153	0.0027	0.0233	0.0085	0.0000	0.0011	0.0000	0.0041	0.0000
i	0.2498	0.0893	0.0100	0.0008	0.0342	0.1194	0.1135	0.0011	0.0250	0.0000	0.0023	0.0002	0.0079
j	0.0000	0.3147	0.0000	0.0000	0.0070	0.0000	0.0000	0.3357	0.0000	0.0000	0.0000	0.0000	0.0000
k	0.0565	0.0198	0.0000	0.0000	0.0085	0.1102	0.0028	0.0028	0.0000	0.0000	0.0000	0.0113	0.0000
l	0.0004	0.0778	0.0041	0.0000	0.0034	0.0389	0.0254	0.0269	0.0056	0.0011	0.0000	0.0819	0.0000
m	0.0042	0.1246	0.0722	0.0000	0.0026	0.0244	0.0005	0.0337	0.0005	0.0000	0.0000	0.0192	0.0000
n	0.0194	0.0528	0.0004	0.0007	0.0011	0.0751	0.1641	0.0124	0.0068	0.0018	0.0002	0.0157	0.0004
o	0.2190	0.0222	0.0292	0.0000	0.1530	0.0357	0.0396	0.0947	0.0334	0.0345	0.0012	0.0041	0.0004
p	0.0006	0.1511	0.0581	0.0000	0.2306	0.0180	0.0287	0.0457	0.0000	0.0000	0.0000	0.0017	0.0000
q	0.0000	0.0000	0.0000	0.0000	0.0000	0.0000	0.0000	1.0000	0.0000	0.0000	0.0000	0.0000	0.0000
r	0.0325	0.1114	0.0055	0.0000	0.0212	0.0655	0.0596	0.0192	0.0142	0.0017	0.0002	0.0306	0.0000
s	0.0021	0.0706	0.0386	0.0009	0.0027	0.0836	0.2483	0.0579	0.0000	0.0039	0.0000	0.0081	0.0000
t	0.0015	0.1229	0.0003	0.0000	0.0479	0.0418	0.0213	0.0195	0.0005	0.0088	0.0000	0.0203	0.0005
u	0.1517	0.0019	0.0386	0.0000	0.1460	0.1221	0.1255	0.0029	0.0014	0.0000	0.0010	0.0014	0.0005
v	0.0000	0.0530	0.0000	0.0000	0.0000	0.0023	0.0000	0.0012	0.0012	0.0000	0.0000	0.0058	0.0000
w	0.0357	0.1292	0.0000	0.0000	0.0106	0.0366	0.0016	0.0000	0.0000	0.0000	0.0000	0.0024	0.0000
x	0.0000	0.0075	0.3507	0.0000	0.0000	0.0000	0.1716	0.0000	0.0000	0.0000	0.0373	0.0000	0.0034
y	0.0172	0.2207	0.0310	0.0000	0.0310	0.1517	0.0172	0.0138	0.0000	0.0103	0.0000	0.0069	0.0034
z	0.0000	0.0506	0.0000	0.0000	0.0000	0.0000	0.0000	0.0127	0.0000	0.0000	0.0000	0.0000	0.0253

Table 1.2 Transition probabilities $p(t \mid s)$ in the English language

Example 1.2: S generates 2-grams with an independent but identical distribution, say $p(s,t), 0 \leq s, t < q$. So

$$Pr_{plain}\{(m_0, m_1, \ldots, m_{2n-1})\} = p(m_0, m_1) \, p(m_2, m_3) \cdots p(m_{2n-2}, m_{2n-1}), n \geq 1.$$

The distribution of 2-grams in English texts can be found in the literature [Kon81, Table 2.3.4]. Of course one can continue like this with tables of the distribution of 3-grams and more. A different approach is given in the following example.

Example 1.3: S generates 1-grams by means of a Markov chain with *transition matrix* $P = (p(t \mid s))_{0 \leq s, t < q}$ and *equilibrium distribution* $p = (p(0), p(1), \ldots, p(q-1))$. So

$$Pr_{plain}\{(m_0, m_1, \ldots, m_{n-1})\} = p(m_0) \, p(m_1 \mid m_0) \, p(m_2 \mid m_1) \cdots p(m_{n-1} \mid m_{n-2}).$$

Here P is a Markov matrix, i.e. a $q \times q$ matrix whose q rows are probability distributions ($p(t \mid s)$ is the probability that the symbol s will be followed by the symbol t), and the vector p is a probability distribution which is the eigenvector of P with eigenvalue 1.
Let P and p be given by Tables 1.2 and 1.3.
We now have the more realistic probabilities

$$Pr_{plain}\{(run)\} = 0.0751 * 0.0192 * 0.1517 \approx 2.19 \times 10^{-4},$$

$$Pr_{plain}\{(urn)\} = 0.0272 * 0.1460 * 0.0325 \approx 1.29 \times 10^{-4},$$

$$Pr_{plain}\{(nru)\} = 0.0814 * 0.0011 * 0.0192 \approx 1.72 \times 10^{-6}.$$

a	0.0723	h	0.0402	o	0.0716	u	0.0272
b	0.0060	i	0.0787	p	0.0161	v	0.0117
c	0.0282	j	0.0006	q	0.0007	w	0.0078
d	0.0483	k	0.0064	r	0.0751	x	0.0030
e	0.1566	l	0.0396	s	0.0715	y	0.0168
f	0.0167	m	0.0236	t	0.0773	z	0.0010
g	0.0216	n	0.0814				

Table 1.3 Equilibrium distribution in English language

Notice that in these three examples the models are *stationary*, i.e., $Pr_{plain}\{(M_j = m_0, M_{j+1} = m_1, \ldots, M_{j+n-1} = m_{n-1})\}$ is independent of j. In the middle of a text one may expect this property to hold, but in other situations this is not the case. Think for instance of the date at the beginning of a letter.

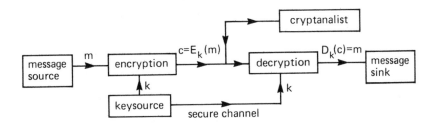

Figure 1.5 Conventional cryptosystem

Now that we have explained what a plaintext source is, we can discuss Figure 1.5.

A *cryptographic transformation* E is a one-to-one mapping of $\mathbb{Z}_q^*$ to $\mathbb{Z}_{q'}^*$. In most practical situations q will be equal to q'. Also often E will map n-grams into n-grams (to avoid data expansion). Let m be some plaintext. Then $c = E(m)$ will be called *ciphertext*. A *cryptosystem* E is a set of cryptographic transformations $E = \{E_k \mid k \in K\}$. The index set K is called the *key space,* and its elements *keys.*

Since E_k is a one-to-one mapping, its inverse must exist. We shall denote it with D_k. Of course the E stands for *encryption* (or *enciphering*) and the D for *decryption* (or *deciphering*). One has

$$\forall_m \forall_{k \in K} [D_k (E_k(m)) = m]. \tag{1.1}$$

If the message source A wants to send data m to message sink B, he will encrypt it with a cryptographic transformation E_k into c. A and B both know the particular choice of the key by means of a *secure channel.* This channel could be a courier, but it could also be that A and B have beforehand agreed on the choice of k. B can decipher c, by computing

$$D_k (c) = D_k (E_k(m)) = m .$$

Normally the same cryptosystem will be used for a long time. It is the frequent changing of the key that has to provide the security of the data. The cryptanalist who is connected to the transmission line can be

i)   *passive* (eavesdropping): The cryptanalist tries to find m (or even better k) from c (and whatever further knowledge he has).

ii) *active* (tampering) : The cryptanalist tries to actively manipulate the data that are being transmitted. For instance he writes his own ciphertext, retransmits old ciphertext, etc..

In general one discerns three levels of cryptanalysis:

ciphertext only attack: Only a piece of ciphertext is known to the cryptanalist (and often the context of the message).

known plaintext attack: A piece of ciphertext with corresponding plaintext is known. If a system is secure against this kind of attack one does not have to destroy deciphered messages.

chosen plaintext attack: The cryptanalist has a by him/her chosen piece of plaintext with corresponding ciphertext. The public key cryptosystems that we shall discuss in later chapters have to be safe against this level of attack.

This concludes our general description of the conventional cryptosystem, as depicted in Figure 1.5. It is now time to discuss some specific cryptosystems.

2 CLASSICAL SYSTEMS

§ 2.1 **Caesar, simple substitution, Vigenère**

In this chapter we shall discuss a number of classical systems. For further reading we refer the interested reader to [Bek82], [Den82], [Kah67], [Kon81] or [Mey82]. One of the oldest cryptosystems is due to Julius Caesar. It shifts each letter in the text cyclicly over k places in the alphabet. In our terminology the *Caesar cipher* is defined by:

$$E_k(i) = (i + k) \underline{\bmod} \ q, 0 \leq i < q,$$
(2.1)

$$E = \{E_k \mid 0 \leq k < q\},$$
(2.2)

where $i \underline{\bmod} n$ denotes the unique integer j, satisfying $j \equiv i \bmod n$ and $0 \leq j < n$. In this case the keyspace K is the set $\{0,1, \ldots ,q-1\}$ and $D_k = E_{q-k}$. For most practical alphabet sizes the cryptanalist can break this system easily by trying all q possible keys. This is called *exhaustive key search*. For instance, when $q = 26$ and we use $\{a,b, \ldots ,z\}$ as alphabet, one only has to check 26 possibilities. In Table 2.1 one can find the cryptanalysis of the ciphertext IYBABZ.

0	IYBABZ	6	OEHGHF	11	TJMLMK	16	YORQRP	21	DTWVWU
1	JZCBCA	7	PFIHIG	12	UKNMNL	17	ZPSRSQ	22	EUXWXV
2	KADCDB	8	QGJIJH	13	VLONOM	18	AQTSTR	23	FVYXYW
3	LBEDEC	9	RHKJKI	14	WMPOPN	19	BRUTUS	24	GWZYZX
4	MCFEFD	10	SILKLJ	15	XNQPQO	20	CSVUVT	25	HXAZAY
5	NDGFGE								

Table 2.1 Cryptanalysis of the Caesar ciphertext IYBABZ

Simple substitution: In this case $K = S_q$, the set of all permutations of $\{0, 1, \ldots, q-1\}$. The cryptosystem E is given by

$$E = \{E_\pi \mid \pi \in S_q\} \tag{2.3}$$

where

$$E_\pi(i) = \pi(i), \quad 0 \le i < q. \tag{2.4}$$

The decryption function D_π is given by $D_\pi = E_{\pi^{-1}}$, since

$$D_\pi(E_\pi(i)) = E_{\pi^{-1}}(E_\pi(i)) = E_{\pi^{-1}}(\pi(i)) = \pi^{-1}(\pi(i)) = i, \ 0 \le i < q.$$

This system does not have the drawback of a small keyspace. Indeed $|S_q| = q!$. For instance with $q = 26$ one has a keyspace of size $26! \approx 4.03 * 10^{26}$. This system however does demonstrate very well that a large keyspace should not fool one into believing that the system is secure! On the contrary. By simply counting the letter frequencies in the ciphertext and comparing these with the letter frequencies in Table 1.2 one very quickly finds the images under π of the most frequent letters in the plaintext. After that it is not difficult at all to fill in the rest. In the following example, we have knowledge of the ciphertext in Table 2.2 and the additional information that the plaintext discusses "bidirectional communication theory". Cryptanalysis will turn out to be very easy.

zhjeo	ndize	hicle	osiol	digic	lmhzq	zolyi	zehdp	zhjeo	ndize
hycdh	hlpvs	uczyc	dhzhj	eondi	zehge	moylk	zhjpm	lhylg	gidiz
gizyd	ppsdo	lylzr	losye	nnmhz	ydize	hicle	osceu	lrloq	lgyoz
vlgic	lneol	flhlo	dpydg	lzhuc	zyciu	eeone	olzhj	eondi	zehge
moylg	zhjpm	lhyll	dycei	clogi	dizgi	zydpp	siclq	zolyi	zehej
iczgz	hjpml	hylzg	lkaol	gglqv	sqzol	yilqi	odhgj	eondi	zehxm
dhizi	zlguc	zycyd	hehps	vlqlo	zrlqz	jiclp	duejy	dmgdp	ziszg
evglo	rlqqz	gizhf	mzgcz	hficl	ldopz	loydm	gljoe	niclp	dilol
jjlyi	zhvze	pefsd	hqgey	zepef	syenn	mhzyd	izehi	cleos	gllng
iecdr	luzql	daapz	ydize	hgqml	ieicl	jdyii	cdipz	rzhfv	lzhfg
dolvs	iclzo	dyize	hggem	oylge	jzhje	ondiz	ehucz	yczhj	pmlhy
lldyc	eiclo	zhdpp	aeggz	vplqz	olyiz	ehgic	laolg	lhiad	aloql
gyzvl	gicly	dglej	vzqzo	lyize	hdpye	nnmhz	ydize	hicle	osdaa
pzlqi	eiclg	eyzdp	vlcdr	zemoe	jneht	lsg..			

Table 2.2 Cipher text obtained with a simple substitution

Assuming that the word "communication" will occur in the plaintext, we look for strings of 13 consecutive letters in the ciphertext, in which letter 1 = letter 8, letter 2 = letter 12, letter 3 = letter 4, letter 6 =letter 13 and letter 7 = letter 11. Indeed we find the string "yennmhzydizeh" three times in the ciphertext. This gives the following information about π.

$$
\begin{array}{cccccccc}
c & o & m & u & n & i & a & t \\
\downarrow & \downarrow & \downarrow & \downarrow & \downarrow & \downarrow & \downarrow & \downarrow \\
y & e & n & m & h & z & d & i
\end{array} \tag{2.5}
$$

Assuming that the word "direction" does also occur in the plaintext, we need to look for strings of the form "$* z * *$ yizeh" in the ciphertext, because of (2.5). It turns out that "qzolyizeh" appears four times, giving:

$$
\begin{array}{ccc}
d & r & e \\
\downarrow & \downarrow & \downarrow \\
q & o & 1
\end{array} \tag{2.6}
$$

If we substitute (2.5) and (2.6) in the ciphertext one easily obtains π completely. For instance the text begins like

in* ormationt* eor* treat* t* eunid... ,

which obviously comes from

information theory treats the unid(irectional)... .

This gives the π-image of the letters f, h, y and s. Continuing like this, one readily obtains π completely

$$
\begin{array}{cccccccccccccccccccccccccc}
a & b & c & d & e & f & g & h & i & j & k & l & m & n & o & p & q & r & s & t & u & v & w & x & y & z \\
\downarrow & \downarrow & \downarrow & \downarrow & \downarrow & \downarrow & \downarrow & \downarrow & \downarrow & \downarrow & \downarrow & \downarrow & \downarrow & \downarrow & \downarrow & \downarrow & \downarrow & \downarrow & \downarrow & \downarrow & \downarrow & \downarrow & \downarrow & \downarrow & \downarrow & \downarrow \\
d & v & y & q & l & j & f & c & z & w & t & p & n & h & e & a & x & o & g & i & m & r & u & k & s & b
\end{array}
$$

The *Vigenère cryptosystem* (named after the Frenchman B. de Vigenère who in 1586 wrote his Traicté des Chiffres, containing a more difficult version of this system) consists of r Caesar ciphers applied periodically. More precisely it is defined by:

$$
E = \{ E_{(k_0, k_1, \ldots, k_{r-1})} \mid (k_0, k_1, \ldots, k_{r-1}) \in K = \mathbb{Z}_q^r \} \tag{2.7}
$$

and

$$
E_{(k_0, k_1, \ldots, k_{r-1})}(m_0, m_1, m_2, \ldots) = (c_0, c_1, c_2, \ldots) \tag{2.8}
$$

with

$$c_l = (m_l + k_{l \bmod r}) \underline{\bmod} \; q. \tag{2.9}$$

Example 2.1: We identify $\{0,1,\ldots,25\}$ with $\{a,b,\ldots,z\}$. The so called *Vigenère Table* (see Table 2.3) is a very helpful tool when encrypting or decrypting. With the key "michael" one gets the following encipherment:

plaintext	a c r y p t o s y s t e m o f t e n i s a c o m p r o m i s e b e t w e e n . . .
key "michael"	m i c h a e l m i c h a e l m i c h a e l m i c h a e l m i c h a e l m i c . . .
ciphertext	m k t f p x z e g u a e q z r b g u i w l o w o w r s x u a g i e x h q m p . . .

Because of the redundancy in the English language one reduces the effective size of the key space tremendously by choosing an existing word as the key. Taking the name of a relative reduces the security of the encipherment more or less to zero.

		a b c d e f g h i j k l m n o p q r s t u v w x y z
0	a	a b c d e f g h i j k l m n o p q r s t u v w x y z
1	b	b c d e f g h i j k l m n o p q r s t u v w x y z a
2	c	c d e f g h i j k l m n o p q r s t u v w x y z a b
3	d	d e f g h i j k l m n o p q r s t u v w x y z a b c
4	e	e f g h i j k l m n o p q r s t u v w x y z a b c d
5	f	f g h i j k l m n o p q r s t u v w x y z a b c d e
6	g	g h i j k l m n o p q r s t u v w x y z a b c d e f
7	h	h i j k l m n o p q r s t u v w x y z a b c d e f g
8	i	i j k l m n o p q r s t u v w x y z a b c d e f g h
9	j	j k l m n o p q r s t u v w x y z a b c d e f g h i
10	k	k l m n o p q r s t u v w x y z a b c d e f g h i j
11	l	l m n o p q r s t u v w x y z a b c d e f g h i j k
12	m	m n o p q r s t u v w x y z a b c d e f g h i j k l
13	n	n o p q r s t u v w x y z a b c d e f g h i j k l m
14	o	o p q r s t u v w x y z a b c d e f g h i j k l m n
15	p	p q r s t u v w x y z a b c d e f g h i j k l m n o
16	q	q r s t u v w x y z a b c d e f g h i j k l m n o p
17	r	r s t u v w x y z a b c d e f g h i j k l m n o p q
18	s	s t u v w x y z a b c d e f g h i j k l m n o p q r
19	t	t u v w x y z a b c d e f g h i j k l m n o p q r s
20	u	u v w x y z a b c d e f g h i j k l m n o p q r s t
21	v	v w x y z a b c d e f g h i j k l m n o p q r s t u
22	w	w x y z a b c d e f g h i j k l m n o p q r s t u v
23	x	x y z a b c d e f g h i j k l m n o p q r s t u v w
24	y	y z a b c d e f g h i j k l m n o p q r s t u v w x
25	z	z a b c d e f g h i j k l m n o p q r s t u v w x y

Table 2.3 Vigenère Table

Instead of using r Caesar ciphers periodically in the Vigenère cryptosystem, one can of course also use r simple substitutions periodically. Such a system is called a *polyalphabetic substitution*. For centuries no one had an effective way of breaking this system, mainly because one did not have a technique of determining the key length r. Once one knows r one can find the r simple substitutions by grouping together the letters $i, r+i, 2r+i, ..., 0 \leq i < r$, and break each of these r simple substitutions individually. In 1863 the Prussian army officer F.W. Kasiski solved the problem of finding the key length r by a technique called *the incidence of coincidences*. We shall now discuss this method.

§ 2.2 The incidence of coincidences

Suppose that $\underline{M_i} = (M_{i,0}, M_{i,1}, \ldots, M_{i,n-1})$, $i = 1,2$ are two sequences of independent and identically distributed random variables on $\mathbb{Z}_q$, say:

$$Pr_{plain} \{M_{i,j} = m\} = p(m), 0 \leq m < q, \ i = 1,2, \ 0 \leq j < n. \tag{2.10}$$

We are interested in the number of coincidences, defined by

$$\kappa[\underline{M_1}, \underline{M_2}] = | \{0 \leq j < n \mid M_{1,j} = M_{2,j}\} | \tag{2.11}$$

Clearly

$$Pr_{plain} \{M_{1,j} = M_{2,j}\} = \sum_m Pr_{plain} \{M_{1,j} = M_{2,j} = m\} = \sum_m p^2(m). \tag{2.12}$$

Let G be a subset of S_q and consider a random variable Π on G with distribution given by

$$Pr_{key} \{\Pi = \pi\} = q(\pi). \tag{2.13}$$

Let $\underline{M_i}, i = 1,2$, be enciphered with the simple substitutions π_1, resp. π_2 in G. Their respective images are $\underline{C_1} = (C_{1,0}, C_{1,1}, \ldots, C_{1,n-1})$ and $\underline{C_2} = (C_{2,0}, C_{2,1}, \ldots, C_{2,n-1})$. Of course for $i = 1,2$ and $0 \leq j < n$

$$Pr_{cipher} \{C_{i,j} = c\} = \sum_{\pi \in G} q(\pi) . p(\pi^{-1}(c)). \tag{2.14}$$

There are two possibilities that we have to consider.

H_0: $\underline{M_1}$ and $\underline{M_2}$ are enciphered with the same simple substitution π (with probability $q(\pi)$).

H_1: $\underline{M_1}$ and $\underline{M_2}$ are enciphered with two independently selected simple substitutions π_1 resp. π_2 (with probability $q(\pi_1)$ resp. $q(\pi_2)$).

Then

$$Pr_{cipher} \{C_{1,j} = C_{2,j} \mid H_0\} = \sum_c Pr_{cipher} \{C_{1,j} = C_{2,j} = c \mid H_0\} =$$

$$\sum_{\pi \in G} q(\pi) \sum_c p^2(\pi^{-1}(c)) = \sum_{\pi \in G} q(\pi) \sum_m p^2(m) = \sum_m p^2(m). \qquad (2.15)$$

while

$$Pr_{cipher}\{C_{1,j} = C_{2,j} \mid H_1\} = \sum_c Pr_{cipher}\{C_{1,j} = C_{2,j} = c \mid H_1\} =$$

$$\sum_{\pi_1, \pi_2 \in G} q(\pi_1) q(\pi_2) \sum_c p(\pi_1^{-1}(c)) p(\pi_2^{-1}(c)) =$$

$$\sum_c \{\sum_{\pi_1 \in G} q(\pi_1) p(\pi_1^{-1}(c))\} \cdot \{\sum_{\pi_2 \in G} q(\pi_2) p(\pi_2^{-1}(c))\} =$$

$$\sum_c \{\sum_{\pi \in G} q(\pi) p(\pi^{-1}(c))\}^2 = \sum_c Pr_{cipher}^2 \{C = c\}. \qquad (2.16)$$

Example 2.2: With Table 1.2 one obtains for the English language the value 0.06875 in formula (2.15). If we take $G = S_{26}$ or G consisting of the 26 Caesar encipherments, and we assume that all elements in G are equally likely, then (2.16) yields the value $1/26 \approx 0.03846$. So the expected value of $\kappa[\underline{C}_1, \underline{C}_2]$ approximately is $0.06875n$ under hypothesis H_0 and $0.03846n$ under hypothesis H_1.

Let us now return to the original problem of determining the key length of a Vigenère cryptosystem.

Let

$$\underline{m} = (m_0, m_1, \ldots\ldots\ldots, m_{n-1}) \quad \text{be the plaintext,}$$

$$\underline{\pi} = (\pi_0, \pi_1, \ldots, \pi_{r-1}) \quad \text{be the key,}$$

$$\underline{c} = (c_0, c_1, \ldots\ldots\ldots, c_{n-1}) \quad \text{be the ciphertext.}$$

So

$$c_i = \pi_{i \bmod r}(m_i).$$

We assume that the m_i's are independent realizations of the random variable M with probability distribution $p(m)$ and that the π_i's are independently selected from G with probability distribution $q(\pi)$. Define

$$\underline{c}^{(s)} = (c_0, c_1, \ldots, c_{n-s-1}) \quad \text{and}$$

$$^{(s)}\underline{c} = (c_s, c_{s+1}, \ldots, c_{n-1}).$$

The expected value of $\kappa[^{(s)}\underline{c}, \underline{c}^{(s)}]$ will turn out to give the clue to determine the value of the key length. Indeed: If $r \mid s$ and $0 \le i < n-s$

$$Pr_{cipher}\{^{(s)}c_i = c_i^{(s)}\} = \sum_c Pr_{cipher}\{c_i = c_{i+s} = c\} =$$

ubsyv	kmhvy	rrtsb	bcrds	ndwrt	shxmb	ufrmx	gabnv	mlrce	weruc
amlyz	brvfw	ivvml	yzwap	spyog	sslec	hbgcu	bsvyc	zqrcw	rmhvc
xgooy	vcyds	pomtq	fpyqk	gbvme	rucad	lcafl	rsuqj	rbhce	qesfc
ehuoq	mdsto	rcdoy	meqqw	aglgo	vggsm	dabbi	gztbb	qyfwb	xwmgf
powgz	tyeil	osrkg	fahuo	vqfog	swruq	nvpwf	vrnmp	qqgss	latgr
mqubs	vyczq	rswcj	deowq	qroih	gdspd	ibffn	xwgzt	bbqyf	wbxus
mbgsx	gqgjr	matqn	xslxm	oohcd	wiohg	rhuop	yicsm	eseoy	sxwug
blwcd	jrnhg	ehvxk	sugus	refvr	oepxw	rbgyg	grpvm	yhuop	yfseo
jdqqg	srzuc	yykwm	bqkrb	ecpss	jaulm	skyia	sgyfw	bxxfq	ceiwc
qafds	fmjrg	mbqoc	zpgoo	gssle	rhoxm	fvroj	dqqgd	lyfzv	fmlsp
rsree	oeofw	fvrsv	yohvy	rqech	bgcec	ssrda	fzkxg	abjrm	atwap
psqbp	oiyov	bdlcd	wakpj	bcfcm	zxsqs	vcohv	yrqfv	rzvce	sadty
bseni	qofvl	iqfvr	meqqc	slmbu	frmxg	abnvg	myahx	mamhv	yrrts
bbcyb	dysib	fcgri	qaqvk	pzqvn	fmmgf	bpqmz	yriwr	iczyr	iqmfy
jyecp	sejsf	betfm	jrlic	zdhdx	mssgr	ipubb	xiamu	rkrbf	vrsvy
ohvyr	qtoio	fcqbb	lwcdj	rnelp	frqmq	fseoh	dafnz	ipucq	yjdut
goily	waexc	etvfi	rkdvm	ejnsu	kzgai	ekpyo	hvfmr	usfci	yfwaq
pcmjv	xkeqb	vdejp	wfzpy	kquev	pooyv	tsevv	xkdaf	zohri	cfder
uggsg	yxhvw	iqqeh	oraqg	gyafu	qudlc	nwqsv	cohvy	ryxqb	wqszw
pkxga	bgrim	dmukw	zqsak	tnxwr	nxfqf	rcyjf	gjovc	fmcsg	yxtbb
xfqgb	mmyxf	rveru	cakrb	mueoi	brceo	eatgr	viafs	qzegd	csdam
ubqsz	gpinv	wricn	vvcmr	lurmi	bzyhg	rwpkx	gabfy	jrtsq	yqgzo
adfct	oisss	dguya	cphlz	mamzi	kpsqg	jsxfd	stkvb	fcgri	aaaze
rgoog	sslci	nxxgf	wrcxf	qprre	tuchb	sdpwp	deraf	vxafu	qudlc
rfroe	afwbx	eqqlc	biqes	qlcrt	ssbic	qbgbs	nkcsd	lcpcz	sryzh
nxmkm	zvckp	qogov	yeqbw	tydsq	gmrth	uowsn	rbwml	mbgyr	cfvrw
ehafo	olyhw	bevyx	wapps	qbpoe	qqlcb	iqesq	lcrts	qsvco	hrnxp
mbfsr	dafzk	xgabt	yiqrf	bwxfq	rbwml	mbgds	rtsfe	fbaav	xelfo
asqyx	huofc	toiss	sdcsr	ipawa	glgov	gridd	srkgr	ucacs	dncgr
eluan	vwyds	nlssf	sdeej	ycfds	dfvrl	ifmjv	yypmz	vxjjg	samie
asfdl	cadcy	wgfsq	svcoh	vyrlm	arvcd	dczdl	ceion	skubn	xxrah
uohmy	wakrr	mbvwe	jfvrn	mafog	yvkar	vpmam	hvyrg	eiacx	ynzrs
xmoqh	bibua	zohgm	hrvcy	rhrbx	fqsfd	ezxwf	rqczh	bpxfq	rbwml
mbgbi	jmhvy	rqtwc	krbov	nxkcp	qbxxg	zibew	jkwak	fmghf	sbuqs
xcxgy	svxxm	fvrri	paabn	mduqn	dmmzk	usgfb	fbfib	fcoox	fqggk
fjqca	o								

Table 2.4 Ciphertext obtained with the Vigenère cryptosystem

$$\sum_{\pi \in G} q(\pi) \sum_c p^2(\pi^{-1}(c)) = \sum_{\pi \in G} q(\pi) \sum_m p^2(m) = \sum_m p^2(m),$$

while if $r \nmid s$ and $0 \le i < n - s$

$$Pr_{cipher}\{^{(s)}c_i = c_i^{(s)}\} = \sum_c Pr_{cipher}\{c_i = c_{i+s} = c\} =$$

$$\sum_{\pi_1, \pi_2 \in G} q(\pi_1) q(\pi_2) \sum_c p(\pi_1^{-1}(c)) p(\pi_2^{-1}(c)) =$$

$$\sum_c \{ \sum_{\pi \in G} q(\pi) p(\pi^{-1}(c)) \}^2 = \sum_c Pr_{cipher}^2 \{ C = c \}.$$

Together this proves the following theorem.

Theorem 2.3 The expected value of $\kappa[^{(s)}\underline{c}, \underline{c}^{(s)}]$ is given by

$$E(\kappa[^{(s)}\underline{c}, \underline{c}^{(s)}]) = \begin{cases} (n-s) \sum_m p^2(m) & \text{if } r \text{ divides } s \quad (2.17) \\ (n-s) \sum_c Pr_{cipher}^2 \{C = c\} & \text{if } r \text{ does not divide } s \quad (2.18) \end{cases}$$

Example 2.4 In Table 2.4 one can find the ciphertext of an English text, that has been encrypted with the Vigenère cryptosystem. In Table 2.5 the values of $\kappa[^{(s)}\underline{c}, \underline{c}^{(s)}]/(n-s)$ are listed for $s = 1, 2, \ldots, 25$. One easily sees (see Example 2.2) that the period is 6. Indeed the key that has been used was the word "monkey", which has 6 letters.

1	0.041	6	0.060	11	0.041	16	0.033	21	0.042
2	0.034	7	0.047	12	0.055	17	0.040	22	0.040
3	0.032	8	0.046	13	0.032	18	0.062	23	0.040
4	0.043	9	0.037	14	0.043	19	0.034	24	0.062
5	0.042	10	0.036	15	0.040	20	0.044	25	0.037

Table 2.5 $\kappa[^{(s)}\underline{c}, \underline{c}^{(s)}]/(n-s)$, $1 \le s \le 25$, for ciphertext in Table 2.4

§ 2.3 Vernam, Playfair, Transpositions, Hagelin, Enigma

The *one-time pad*, also called the *Vernam* cipher (after the American A.T.& T. employee G.S. Vernam, who introduced this system in 1917), is a Vigenère cipher with key length r equal to the length n of the plaintext. Also the key must be chosen in a completely random way. In this way the system is unconditionally secure, as is intuitively clear and shall be proved in Chapter 4. The "hot line" between Washington and Moscow uses this system. The major drawback of this system is the length of the key, which makes this system not very practical for most applications.

We shall now briefly discuss a few more cryptosystems, without going very deep into their structure.

The *Playfair* cipher (1854, named after the Englishman L. Playfair) was used by the British in World War I. It operates on 2-grams. First of all, one identifies the letters i and j. The 25 remaining letters are put rowwise in a 5×5 matrix K, beginning with the letters of a keyword. If a letter occurs more than once in the keyword, it is only used once. The missing letters are put into K in their natural order. For instance the keyword "Hieronymus" gives rise to:

$$
K = \begin{matrix}
h & i & e & r & o \\
n & y & m & u & s \\
a & b & c & d & f \\
g & k & l & p & q \\
t & v & w & x & z
\end{matrix}
$$

The 2-gram $(x,y) = (K_{i,j}, K_{m,n})$ with $x \neq y$, will be enciphered into

$(K_{i,n}, K_{m,j})$ if $i \neq m$ and $j \neq n$,

$(K_{i,j+1}, K_{i,n+1})$ if $i = m$ and $j \neq n$,

$(K_{i+1,j}, K_{m+1,j})$ if $i \neq m$ and $j = n$,

where the indices are taken modulo 5. If $x = y$ one inserts the letter q and enciphers the text $\cdots xqy \cdots$.

A completely different way of enciphering is named *transposition*. This system breaks the text up into blocks of fixed length, say n, and applies a fixed permutation σ to the coordinates. For instance with $n = 5$ and $\sigma = (1,4,5,2,3)$ the plaintext

       crypt   **ograp**   **hical**   ...

will become

       ytrcp   **rpgoa**   **cliha**   ...

Often the permutation is of a geometrical nature, as is the case with the so-called *column transposition*. The plaintext is written rowwise in a matrix of given size, but will be read out columnwise in a specific order depending on a keyword. For instance after giving the letters $a, b, \ldots, z$ the numbers $1, 2, \ldots, 26$, the keyword "right" will dictate you to first read out column 3 (being the alphabetically first of the 5 letters in "right"), followed by column 4, 2, 1 and 5. So the plaintext

       "computing science has had very little influence on computing practice"

when encrypted with a 5×5 matrix and keyword "right" will first be written as

```
4 3 1 2 5      4 3 1 2 5      4 3 1 2 5
c o m p u      y l i t t      n g p r a
t i n g s      l e i n f      c t i c e
c i e n c      l u e n c      . . .
e h a s h      e o n c o
a d v e r      m p u t i
```

and then read out like

> "mneav pgnse oiihd ctcea uschr iienu tnnct leuop yllem tfcoi ..".

Since transpositions do not change letterfrequencies, but destroy dependencies between consecutive letters in the plaintext, while Vigenère etc. do the oppposite, one often combines systems like these. Such a combined system is called a *product cipher*. Shannon used the words *confusion* and *diffusion* in this context.

Ciphersystems that encrypt symbol for symbol are often called *streamciphers*, while if they encrypt blocks of a fixed number of symbols simultaneously, they are called *block ciphers*. From a mathematical point of view this distinction is not very important.

During World War II both sides used so called *rotor machines* for their encryption. Several variations of the machines described below have been in use at that time. We shall only give a rough idea of one variation of each.

The *Hagelin*, invented by the Swede B. Hagelin and used by the U.S. Army, has 6 rotors with 26, resp. 25, 23, 21, 19, 17 pins. Each of these pins can be put into an active or passive position. After encryption of a letter (depending on the setting of these pins), the 6 rotors all turn one position. So after 26 encryptions the first rotor is back in its original position. For the sixth rotor this takes only 17 encryptions. Since the numbers of pins on the rotors are coprime, the Hagelin can be viewed as a mechanical Vigenère cryptosystem with period $26 \times 25 \times 23 \times 21 \times 19 \times 17 = 101,405,850$. We refer the reader who is interested in the cryptanalysis of the Hagelin to [Bek82, § 2.3].

The electro-mechanical *Enigma*, used by Germany and Japan, was invented by A. Scherbius in 1923. It consists of three rotors and a reflector. See Figure 2.6. When punching in a letter an electric current will enter the first rotor at a place corresponding with that letter, but will leave it somewhere else depending on the internal wiring of that rotor. The second and third rotor do the same, but have a different wiring. The reflector returns the current at a different place and the current will go through rotors 1,2 and 3 again but in reversed order. The current will light up a letter, which gives the encryption of the original letter. Simultaneously the first rotor will turn one position. After 26 rotations of the first rotor the second one will turn one position. When the second rotor has made a full cycle, the third rotor will rotate over one position. The key consists of i) the choice of the rotors ii) their initial position and iii) a fixed initial permutation of the alphabet. For an idea about the cryptanalysis of the Enigma the reader is referred to [Kon81, Chapter 5].

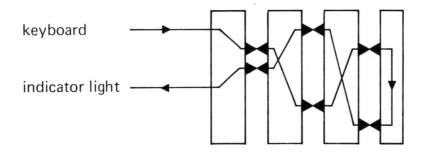

Figure 2.6 Schematic description of the Enigma rotors

Problems

1. The following ciphertext about president Kennedy has been made with a simple substitution. What is the corresponding plaintext?

    rgjjg mvkto tzpgt stbgp catjw pgocm gjs

2. Decrypt the following ciphertext, which is made with the Playfair cipher and the key "hieronymus" (as on page 15).

    erohh mfimf ienfa bsesn pdwar gbhah ro

3. Encrypt the following plaintext using the Vigenère system with key "vigenere":

    who is afraid for virginia wolf

3 SHIFT REGISTER SEQUENCES

§ 3.1 Introduction

During and after World War II the use of logical circuits made completely electronic cryptosystems possible. These turned out to be very practical in the sense of being easy to implement and very fast. The analysis of their security is not so easy! Working with logical circuits often leads to the alphabet $\{0,1\}$. There are only two permutations of the symbols 0 and 1. One interchanges the two symbols. This can also be described by adding 1 (modulo 2) to the two elements. The other leaves the two invariant, which is the same as adding 0 (modulo 2) to these two elements. Since the Vernam cipher is unconditionally secure but not very practical, it is only natural that people came up with the following scheme.

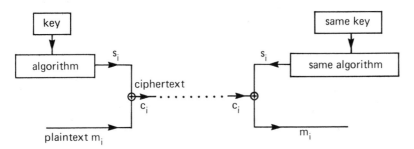

Figure 3.1 A binary cryptosystem with pseudo-random $\{s_i\}_{i \geq 0}$-sequence

Of course one would like the sequence $\{s_i\}_{i \geq 0}$ to be random, but with a finite state machine and a deterministic algorithm one cannot generate a random sequence. Indeed one will generate a sequence, which is ultimately periodic. This makes the scheme a special case of the Vigenère cryptosystem. On the other hand, one can try to generate sequences that appear to be random, have long periods and have the right cryptographic properties. Good reference books for this theory are [Bek82], [Gol67] and [Rue86].

In [Gol67] S.W. Golomb formulated three postulates that a binary, periodic sequence $\{s_i\}_{i\geq0}$ should satisfy to be called *pseudo-random*. Before we can give these, we have to introduce some terminology. Let $\{s_i\}_{i\geq0}$ be a *periodic*, binary sequence, say with *period* p. This means that p is the smallest positive integer for which

$$s_{i+p} = s_i \qquad\qquad\qquad (3.1)$$

for all $i \geq 0$.

A *run* of length k starts at moment t if $s_{t-1} \neq s_t = s_{t+1} = ... = s_{t+k-1} \neq s_{t+k}$.
One makes the following distinction:

$$\leftarrow\ k\ \rightarrow$$

a *block* of length k: 011.........110
a *gap* of length k: 100.........001

The *autocorrelation* $AC(k)$ of a periodic sequence $\{s_i\}_{i\geq0}$ with period p is defined by:

$$AC(k) = \frac{A-D}{p}, \qquad\qquad\qquad (3.2)$$

where A and D are the number of agreements resp. disagreements over a full period between $\{s_i\}_{i\geq0}$ and $\{s_{i+k}\}_{i\geq0}$, which is $\{s_i\}_{i\geq0}$ shifted over k positions to the left. So

$$A = |\ \{0 \leq i < p\ |\ s_i = s_{i+k}\}\ | \qquad\qquad\qquad (3.3)$$

$$D = |\ \{0 \leq i < p\ |\ s_i \neq s_{i+k}\}\ |\ . \qquad\qquad\qquad (3.4)$$

If k is a multiple of p one speaks of the *in-phase* autocorrelation. It always has value 1. If p does not divide k, one speaks of the *out-of-phase* autocorrelation. Its value lies between -1 and 1.

Golombs Randomness Postulates

G1:   The number of zeros and the number of ones are as equal as possible per period, i.e. $p/2$ if p is even and $(p \pm 1)/2$ if p is odd.

G2:   Half of the runs in a cycle have length 1, one quarter of the runs have length 2, one eighth of the runs have length 3, and so forth. Moreover half of the runs of a certain length are gaps, the other half are blocks.

G3: The out-of-phase autocorrelation $AC(k)$ has the same value for all k.

G1 states that zeros and ones occur with roughly the same probability. G2 implies that after 01 the symbol 0 has about the same probability as the symbol 1, etc.. So G2 says that certain n-grams occur with the right probabilities. The interpretation of G3 is more difficult. It does say that counting the number of agreements between a sequence and a shifted version of that sequence does not give any information about the period of that sequence, unless one shifts over a multiple of the period. Compare this with Theorem 2.3, where this comparison gives the length of the key used in a Vigenère cipher. In cryptographical applications p will be too large for such an approach. The value of the constant in G3 follows from G1 and G3, as we shall now prove.

Lemma 3.1 Let $\{s_i\}_{i \geq 0}$ be a binary sequence with period p, which satisfies the randomness postulates.

Then for $p \nmid k$

$$AC(k) = \begin{cases} -1/(p-1) & \text{if } p \text{ is even} & (3.5) \\ -1/p & \text{if } p \text{ is odd.} & (3.6) \end{cases}$$

Proof: Consider a $p \times p$ cyclic matrix with top row $s_0, s_1, \ldots, s_{p-1}$. We count in two different ways the sum over $1 \leq i \leq p-1$ of the agreements minus the disagreements between the top row and row i. Counting rowwise we get by G3 for each $1 \leq i \leq p-1$ the same value $p \cdot AC(k)$. This gives a total value of $p(p-1) \cdot AC(k)$. On the other hand, when counting columnwise, one gets by G1 the contribution $(p/2-1)-p/2 = -1$ for each column when p is even. This gives $-p$ as value of the summation. Equating the two values yields (3.5). If p is odd, one gets for $(p+1)/2$ columns the contribution $(p-1)/2 - (p-1)/2$, which is 0, and for $(p-1)/2$ columns the contribution $(p-3)/2 - (p+1)/2$, which is -2. Hence one obtains the value $-(p-1)$ for the summation. Putting this equal to $p(p-1) \cdot AC(k)$ yields (3.6). □

The well known χ^2-test and the spectral test [Cov67] yield ways to test the pseudo-randomness properties of a given sequence. We shall not discuss these methods here. The interested reader is referred to [Gol67, chapter IV], and [Knu69, chapter 3].

There are also properties of a cryptographic nature which the sequence $\{s_i\}_{i \geq 0}$ should satisfy.

C1: The period p of $\{s_i\}_{i \geq 0}$ has to be taken very large (about the order of magnitude of 10^{50}).

C2: The sequence $\{s_i\}_{i \geq 0}$ should be easy to generate.

C3: Knowledge of part of the plaintext with corresponding ciphertext should not enable a cryptanalist to generate the whole $\{s_i\}_{i \geq 0}$-sequence.

§ 3.2 Linear feedback shift registers

Feedback shift registers are very fast generators of binary sequences. Their general form is depicted in Figure 3.2.

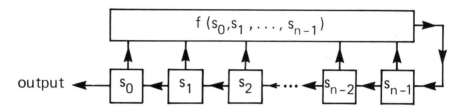

Figure 3.2 General feedback shift register.

A feedback shift register of *length* n consists of n memory cells, which together form the *state* $(s_0, s, \ldots, s_{n-1})$ of the shift register. The function f is a mapping of $\{0,1\}^n$ in $\{0,1\}$ and is called the *feedback function*. Since f is a boolean function it can easily be made with logical switches. After the first time unit the shift register will output s_0 and go to the state $(s_1, s_2, \ldots, s_n)$, where $s_n = f(s_0, s_1, \ldots, s_{n-1})$. Continuing in this way the shift register will generate an infinite sequence $\{s_i\}_{i \geq 0}$. We shall first study the special case that f is a linear function, say:

$$f(s_0, s_1, \ldots, s_{n-1}) = c_0 s_0 + c_1 s_1 + \cdots + c_{n-1} s_{n-1}, \tag{3.7}$$

where $c_i \in \{0,1\}$, $0 \leq i \leq n-1$, and all the additions are taken modulo 2. The output sequence $\{s_i\}_{i \geq 0}$ can now be described by the starting values s_i, $0 \leq i \leq n-1$, and the recurrence relation:

$$s_{k+n} = \sum_{i=0}^{n-1} c_i s_{k+i}, \quad k \geq 0, \tag{3.8a}$$

or, equivalently

$$\sum_{i=0}^{n} c_i s_{k+i} = 0, \quad k \geq 0, \tag{3.8b}$$

where $c_n = 1$ by definition.

Let $\underline{s}^{(i)}$ denote the state at time i, i.e. $\underline{s}^{(i)} = (s_i, s_{i+1}, \ldots, s_{i+n-1})$. Then similar to (3.8a) one has

$$\underline{s}^{(k+n)} = \sum_{i=0}^{n-1} c_i \underline{s}^{(k+i)}, \quad k \geq 0 . \tag{3.8c}$$

The general picture of a *linear feedback shift register*, which name we shall shorten to LFSR, is depicted in Figure 3.3.

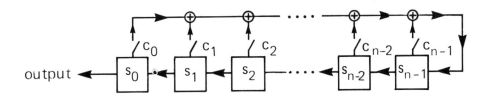

Figure 3.3 General linear feedback shift register (LFSR)

The coefficients c_i are called the *feedback coefficients*. If $c_i = 0$ then the corresponding switch in Figure 3.3 is open, while if $c_i = 1$ this switch is closed. We shall always assume that $c_0 = 1$, because otherwise the output sequence $\{s_i\}_{i \geq 0}$ is just a delayed version of a sequence, generated by a LFSR with its c_0 equal to 1. As a consequence any state of the LFSR not only has a unique successor state, as is natural, but also has a unique predecessor. Later we shall prove this property in a more general situation.

Example 3.2 With $n = 4$, $c_0 = c_1 = 1$, $c_2 = c_3 = 0$ and starting state $(1,0,0,0)$ one gets the following LFSR and list of consecutive states:

time	s_0	s_1	s_2	s_3
0	1	0	0	0
1	0	0	0	1
2	0	0	1	0
3	0	1	0	0
4	1	0	0	1
5	0	0	1	1
6	0	1	1	0
7	1	1	0	1
8	1	0	1	0
9	0	1	0	1
10	1	0	1	1
11	0	1	1	1
12	1	1	1	1
13	1	1	1	0
14	1	1	0	0
15	1	0	0	0 , etc. .

Note that the state at $t = 15$ is identical to the state at $t = 0$, so the output sequence $\{s_i\}_{i \geq 0}$ has period 15.

Since there are precisely 2^n different states in a LFSR of length n and the all-zero state always goes over into itself, one can conclude that the period of $\{s_i\}_{i \geq 0}$ is at most $2^n - 1$.

Definition 3.3 A *PN-sequence* or *pseudo-noise sequence* is an output sequence of an n-stage LFSR that has period $2^n - 1$.

If an n-stage LFSR does not run cyclically through all $2^n - 1$ non-zero states, it certainly does not generate a PN-sequence. As a consequence we have the following theorem.

Theorem 3.4 An n-stage LFSR that generates a PN sequence $\{s_i\}_{i \geq 0}$, runs cyclically through all $2^n - 1$ non-zero states. Any non-zero output sequence of this LFSR is a shift of $\{s_i\}_{i \geq 0}$.

We want to know which LFSR's generate PN-sequences. To this end we associate with an LFSR with feedback coefficients $c_0, c_1, \ldots, c_{n-1}$ a so called *characteristic polynomial* $f(x)$ defined by:

$$f(x) = c_0 + c_1 x + c_2 x^2 + \cdots + c_{n-1} x^{n-1} + x^n = \sum_{i=0}^{n} c_i x^i, \tag{3.9}$$

where $c_n = 1$ by definition and $c_0 = 1$ by assumption.

Definition 3.5 Let f be given by (3.9). Then

$$\Omega(f) = \{ \{s_i\}_{i \geq 0} \mid \{s_i\}_{i \geq 0} \text{ satisfies } (3.8a) \} . \tag{3.10}$$

In words $\Omega(f)$ is the set of all output sequences of the LFSR with characteristic polynomial $f(x)$.

Lemma 3.6 Let f be the characteristic polynomial of an n-stage LFSR. Then $\Omega(f)$ is a binary vector space of dimension n.

Proof: Since (3.8a) is a linear recurrence relation, $\Omega(f)$ is obviously a vectorspace. Also each $\{s_i\}_{i \geq 0}$ in $\Omega(f)$ is uniquely determined by $s_0, s_1, \ldots, s_{n-1}$, so the dimension of $\Omega(f)$ is at most n. On the other hand the n different sequences starting with

$$\overset{\longleftarrow \ i \ \longrightarrow}{} \overset{\longleftarrow \ n-i-1 \ \longrightarrow}{}$$
$$00 \ldots \ldots \ldots 0100 \ldots \ldots \ldots 0 \quad ,$$

$0 \leq i \leq n-1$, are clearly independent. So the dimension of $\Omega(f)$ is at least n.   []

Let f be a polynomial of degree n, say $f(x) = \sum_{i=0}^{n} c_i x^i$. Then the *reciprocal polynomial* of $f(x)$ is defined by

$$f^*(x) = x^n f(1/x) = c_0 x^n + c_1 x^{n-1} + \cdots c_{n-1} x + c_n . \tag{3.11}$$

With a sequence $\{s_i\}_{i\geq 0}$ we associate the power series (also called *generating function*)

$$S(x) = \sum_{i=0}^{\infty} s_i x^i . \tag{3.12}$$

Instead of writing $\{s_i\}_{i\geq 0} \in \Omega(f)$, we shall also use the notation $S(x) \in \Omega(f)$. We know that $S(x)$ is uniquely determined by $s_0, s_1, \ldots, s_{n-1}$ and $f(x)$. We shall now make this dependency more explicit.

Theorem 3.7 Let $\{s_i\}_{i\geq 0} \in \Omega(f)$, with f given by (3.9). Let $S(x)$ be the generating function of $\{s_i\}_{i\geq 0}$. Then

$$S(x) = \frac{\tau(x)}{f^*(x)} , \tag{3.13}$$

where

$$\tau(x) = \sum_{j=0}^{n-1} \left(\sum_{l=0}^{j} c_{n-l} s_{j-l} \right) x^j . \tag{3.14}$$

Proof: By (3.11), (3.12) and (3.8b)

$$S(x)f^*(x) = \left(\sum_{k=0}^{\infty} s_k x^k \right) \left(\sum_{l=0}^{n} c_{n-l} x^l \right) =$$

$$= \sum_{j=0}^{\infty} \left(\sum_{l=0}^{\min\{j,n\}} c_{n-l} s_{j-l} \right) x^j =$$

$$= \sum_{j=0}^{n-1} \sum_{l=0}^{j} c_{n-l} s_{j-l} x^j + \sum_{j\geq n} \sum_{l=0}^{n} c_{n-l} s_{j-l} x^j =$$

$$= \tau(x) + \sum_{j\geq n} \left(\sum_{i=0}^{n} c_i s_{(j-n)+i} \right) x^j = \tau(x) . \qquad \Box$$

Corollary 3.8

$$\Omega(f) = \{ \tau(x)/f^*(x) \mid \text{degree of } \tau(x) < n \} . \tag{3.15}$$

Proof: From Theorem 3.7 we know that each member of $\Omega(f)$ can be written uniquely as $\tau(x)/f^*(x)$ with degree $(\tau(x)) < n$. On the other hand, $\Omega(f)$ has cardinality 2^n and there are exactly 2^n binary polynomials $\tau(x)$ of degree $< n$. $\qquad \Box$

It is now easy to prove the following lemma.

Lemma 3.9 Let $\{s_i\}_{i\geq 0} \in \Omega(f)$ and $\{t_i\}_{i\geq 0} \in \Omega(g)$. Let $lcm\ [f,g]$ denote the least common multiple of f and g. Then

$$\{s_i + t_i\}_{i\geq 0} \in \Omega(lcm\ [f,g]).$$

Proof: Write $h = lcm\ [f,g]$, $h = u \cdot f$ and $h = v \cdot g$.

It follows from Corollary 3.8 that $S(x) = \alpha(x)/f^*(x)$, $T(x) = \beta(x)/g^*(x)$, where degree $(\alpha(x)) <$ degree $(f(x))$ and degree $(\beta(x)) <$ degree $(g(x))$. Since $S(x) + T(x) = \alpha(x)/f^*(x) + \beta(x)/g^*(x) = = (\alpha(x)u^*(x) + \beta(x)v^*(x))/h^*(x)$ and $\alpha(x)u^*(x)$ as well as $\beta(x)v^*(x)$ have degree less than degree $(h(x))$, it follows that $S(x) + T(x) \in \Omega(h)$.   []

A polynomial $f(x)$ of degree n is called *reducible* if there exist polynomials $a(x)$ and $b(x)$, both of degree $< n$, such that $f(x) = a(x) \cdot b(x)$. Otherwise $f(x)$ is called *irreducible*. We remind the reader that in this chapter all polynomials will be binary, i.e. with coefficients in $\{0,1\}$, the finite field with two elements. From the theory of finite fields (see Appendix B) we recall the following facts.

Fact 3.10   For each polynomial $f(x)$, $f(0) = 1$, there exists a positive integer m such that $f(x)$ divides $x^m - 1$ (instead of $x^m - 1$ one may also write $x^m + 1$, because these two are the same over $GF(2)$). The smallest value of such an integer m is called the *period* of $f(x)$.

Fact 3.11   If $f(x)$ is irreducible and of degree n then the period of $f(x)$ divides $2^n - 1$. If the period of an irreducible polynomial $f(x)$ is equal to $2^n - 1$ then x is called a *primitive element* in $GF(2)[x]/(f(x))$, and $f(x)$ is called a *primitive polynomial*.

Fact 3.12   The number of primitive polynomials of degree n is given by $\phi(2^n - 1)/n$, where *Euler's totient function* ϕ is given by

$$\phi(m) = |\ \{1 \leq i \leq m \mid gcd(i,m) = 1\}\ | = m \prod_{\substack{p \mid m \\ p\ prime}} (1 - \frac{1}{p}). \qquad (3.16)$$

We need three lemmas before we can characterize the polynomials $f(x)$ that generate PN-sequences.

Lemma 3.13 Let $f(x)$ have period m and degree n. Let $\{s_i\}_{i\geq 0} \in \Omega(f)$. Then $\{s_i\}_{i\geq 0}$ has a period dividing m.

Proof: By assumption there exists a polynomial $g(x)$ such that $x^m + 1 = f(x) \cdot g(x)$ and degree $(g(x)) = m - n$. Taking the reciprocal of both sides, yields $x^m + 1 = f^*(x) \cdot g^*(x)$. By Theorem 3.7 there exists a polynomial $\tau(x)$ of degree $< n$ such that

$$S(x) = \frac{\tau(x)}{f^*(x)} = \frac{\tau(x) \cdot g^*(x)}{1 + x^m} = \tau(x) \cdot g^*(x) \cdot \{1 + x^m + x^{2m} + \cdots\}.$$

Since degree $(g^*(x)) = m - n$, it follows that degree $(\tau(x) \cdot g^*(x)) < m$. So $S(x)$ has a period dividing m. $\qquad \Box$

Lemma 3.14 Let $f(x)$ have period m, degree n and be irreducible. Let $\{s_i\}_{i \geq 0} \in \Omega(f)$. Then $\{s_i\}_{i \geq 0}$ has period equal to m.

Proof: Let $\{s_i\}_{i \geq 0}$ have period p. By Lemma 3.13 p divides m. It follows that $S(x) = u(x)/(1 + x^p)$ for some $u(x)$ of degree $< p$, while on the other hand $S(x) = \tau(x)/f^*(x)$ by (3.13). Equating these two expressions yields

$$(1 + x^p) \cdot \tau(x) = u(x) \cdot f^*(x),$$

and thus

$$(1 + x^p) \cdot \tau^*(x) = u^*(x) \cdot f(x).$$

Since $f(x)$ is irreducible and degree $(\tau^*(x)) < n$, it follows that $f(x)$ divides $(x^p + 1)$. But $f(x)$ has period m, so m divides p. We conclude that $p = m$. $\qquad \Box$

Lemma 3.15 Let $f(x)$ be a polynomial of degree n. Let $\{s_i\}_{i \geq 0} \in \Omega(f)$ be a PN-sequence. Then f is irreducible.

Proof: Suppose that $f(x) = f_1(x) \cdot f_2(x)$, with $f_1(x)$ irreducible and of degree $n_1 > 0$. By Corollary 3.8 $1/f_1^*(x) \in \Omega(f_1)$, so the period of $1/f_1^*(x)$ divides $2^{n_1} - 1$ by Lemma 3.13 and Fact 3.11. On the other hand that same $1/f_1^*(x) = f_2^*(x)/f^*(x) \in \Omega(f)$, so by Theorem 3.4 $1/f_1^*(x)$ is a shift of $\{s_i\}_{i \geq 0}$ and thus its period is $2^n - 1$. This is only possible if $n = n_1$, i.e. $f(x) = f_1(x)$. $\qquad \Box$

Lemmas 3.14 and 3.15 together prove the following theorem.

Theorem 3.16 An output sequence of an LFSR with characteristic polynomial f is a PN-sequence if and only if f is primitive.

Corollary 3.17 There are $\phi(2^n - 1)/n$ different n-stage LFSR's, that generate PN-sequences.

We shall now investigate to which extent PN-sequences meet Golomb's randomness postulates G1-G3. After that, we check the cryptographic requirements C1-C3.

Ad G1:   By Theorem 3.4 each non-zero state occurs exactly once per period. The leftmost bit of each state will be the next output bit. So $\# 1's$ per period is 2^{n-1} and $\# 0's$ per period is $2^{n-1} - 1$.

Ad G2:   There are 2^{n-k-2} states whose leftmost $k+2$ coordinates are of the form $011 \ldots 10$ resp. $100 \ldots 01$. So gaps and blocks of the length $k \le n-2$ occur exactly 2^{n-k-2} times per period. The state $(0,1,1,\ldots,1)$ occurs exactly once. Its successor is $(1,1,\ldots,1)$, which in turn is followed by state $(1,1,\ldots,1,0)$. So there is no block of length $n-1$ and one block of length n. Similarly there is one gap of length $n-1$ and no gap of length n.

Ad G3:   With $\{s_i\}_{i \ge 0} \in \Omega(f)$ also $\{s_{i+k}\}_{i \ge 0} \in \Omega(f)$ by Theorem 3.4. The linearity of $\Omega(f)$ implies that also $\{s_i + s_{i+k}\}_{i \ge 0} \in \Omega(f)$. The number of agreements per period between $\{s_i\}_{i \ge 0}$ and $\{s_{i+k}\}_{i \ge 0}$ equals the number of zeros in one period of $\{s_i + s_{i+k}\}_{i \ge 0}$, which is $2^{n-1} - 1$ by Theorem 3.4 and G1. Similarly the number of disagreements is 2^{n-1}. So the out-of-phase autocorrelation $AC(k)$ is $-1/(2^n - 1)$, for all $1 \le k < 2^n - 1$.

We conclude that PN-sequences meet Golomb's randomness postulates in a most satisfactory way. Let us now check C1-C3.

Ad C1:   Since the period of a PN-sequence generated by an n-stage LFSR is $2^n - 1$, one can easily get sufficient large periods. For instance with $n = 166$ the period is already about 10^{50}.

Ad C2:   LFSR's are extremely simple to implement.

Ad C3:   PN-sequences are very unsafe!! Indeed knowledge of $2n$ consecutive bits, say $s_k, s_{k+1}, \ldots, s_{k+2n-1}$, enables the cryptanalist to determine the feedback coefficients c_i, $0 \le i \le n-1$, uniquely and thus the whole $\{s_i\}_{i \ge 0}$ -sequence. This follows from the matrix equation:

$$\begin{bmatrix} s_k & s_{k+1} & \cdots & s_{k+n-1} \\ s_{k+1} & s_{k+2} & \cdots & s_{k+n} \\ \cdot & \cdot & \cdots & \cdot \\ \cdot & \cdot & \cdots & \cdot \\ \cdot & \cdot & \cdots & \cdot \\ s_{k+n-1} & s_{k+n} & \cdots & s_{k+2n-2} \end{bmatrix} \begin{bmatrix} c_0 \\ c_1 \\ \cdot \\ \cdot \\ \cdot \\ c_{n-1} \end{bmatrix} = \begin{bmatrix} s_{k+n} \\ s_{k+n+1} \\ \cdot \\ \cdot \\ \cdot \\ s_{k+2n-1} \end{bmatrix} \qquad (3.17)$$

Clearly the state $(0,0,\ldots,0,1)$ and its $n-1$ successor states are independent. It follows with an induction argument from (3.8c) that any n consecutive states are independent. This makes (3.17) into a system of n independent equations with the n unknowns $c_i, 0 \le i \le n-1$. So the feedback coefficients can easily be determined.

Remark 3.18 If only a string of $2n-1$ consecutive bits of a PN-sequence is known, the feedback coefficients are not necessarily unique, as follows from the example $n = 4$ and the subsequence 1101011.

Since sequences generated by LFSR's fail to meet requirement C3, the next step will be to study nonlinear shift registers. However since one knows so much about PN-sequences, it is quite natural that one tries to combine LFSR's in a non-linear way in order to get pseudo-random sequences with the right cryptographic properties.

§ 3.3 Non-linear algorithms

As already remarked in § 3.1 any deterministic algorithm in a finite state machine will generate a sequence $\{s_i\}_{i\geq0}$, which is ultimately periodic, say with period p. This means that except for a beginning part, $\{s_i\}_{i\geq0}$ will be generated in a trivial way by the LFSR with characteristic polynomial $1 + x^p$. So the sequence $\{s_i\}_{i\geq0}$ which was possibly made in a non-linear way, can also be made by a LFSR (except for a finite beginning part). If this beginning part is non empty, not every state has a unique predecessor and the output sequence certainly will not have maximal period. We shall come back to this problem later. Here we shall assume that the output sequence is periodic from the start on. The discussion above justifies the following definition.

Definition 3.19 The *linear equivalence* of a periodic sequence $\{s_i\}_{i\geq0}$ is the length n of the smallest LFSR that can generate $\{s_i\}_{i\geq0}$.

The following two lemmas are needed to prove explicit statements about the linear equivalence of periodic sequences.

Lemma 3.20 Let h and f be the characteristic polynomials of an m-stage LFSR resp. n-stage LFSR. Then $\Omega(h) \subset \Omega(f)$ iff $h \mid f$.

Proof:
$\Rightarrow$ Since $1/h^* \in \Omega(h) \subset \Omega(f)$, it follows from Corollary 3.8 that for some polynomial $\tau(x)$ of degree $< n$ one has $1/h^*(x) = \tau(x)/f^*(x)$. We conclude that $f^*(x) = h^*(x) \cdot \tau(x)$ and thus that $f(x) = h(x) \cdot \tau^*(x)$ i.e. $h \mid f$.
$\Leftarrow$ Writing $f(x) = a(x) \cdot h(x)$ with degree $(a(x)) = n - m$, one has by Corollary 3.8 that

$$\Omega(h) = \{ \ \frac{\tau(x)}{h^*(x)} \mid \text{degree } (\tau(x)) < m \ \} =$$

$$\{ \ \frac{a^*(x) \cdot \tau(x)}{a^*(x) \cdot h^*(x)} \mid \text{degree } (\tau(x)) < m \ \} =$$

$$\{ \ \frac{a^*(x) \cdot \tau(x)}{f^*(x)} \mid \text{degree } (a^*(x) \cdot \tau(x)) < n \ \} \subset$$

$$\{ \ \frac{\delta(x)}{f^*(x)} \mid \text{degree } (\delta(x)) < n \ \} = \Omega(f) .$$

[]

Let $\{s_i\}_{i\geq0} \in \Omega(f)$ for some f and suppose that one wants to find the polynomial h of smallest degree such that $\{s_i\}_{i\geq0} \in \Omega(h)$. Then Lemma 3.20 suggests checking the divisors of f. That this is sufficient will be proved later. The next lemma says when one does not need to check the divisors of f.

Lemma 3.21 Let $\{s_i\}_{i\geq0} \in \Omega(f)$ and $S(x) = \tau(x)/f^*(x)$. Then

$$\exists_{h\,|\,f,\,h\neq f}\,[\{s_i\}_{i\geq0} \in \Omega(h)] \quad\Leftrightarrow\quad gcd\,(\tau(x),f^*(x)) \neq 1.$$

Proof: Let $d(x)\,|\,gcd\,(\tau(x),f^*(x))$ with degree $(d(x)) = d \geq 1$.

Then $S(x) = \dfrac{\tau(x)/d(x)}{f^*(x)/d(x)}$, so $\{s_i\}_{i\geq0} \in \Omega(f/d^*)$. It follows that there exists a proper divisor h of f with $\{s_i\}_{i\geq0} \in \Omega(h)$. The proof in the reverse direction goes exactly the same. ⬚

Theorem 3.22 Let $\{s_i\}_{i\geq0}$ be a binary, periodic sequence, say with period p. Let $S^{(p)}(x) = s_0 + s_1 x + \cdots + s_{p-1}x^{p-1}$. Then there exists a unique polynomial $m(x)$ with

i) $\{s_i\}_{i\geq0} \in \Omega(m)$,

ii) $\forall_h\,[\{s_i\}_{i\geq0} \in \Omega(h) \Rightarrow m\,|\,h]$.

The reciprocal of this polynomial m is given by

$$m^*(x) = \frac{1 + x^p}{gcd\,(S^{(p)}(x), 1+x^p)}\,, \tag{3.18}$$

and $m(x)$ is called the *minimal characteristic polynomial* of $\{s_i\}_{i\geq0}$.

Proof: Let $\{s_i\}_{i\geq0} \in \Omega(f)$, but $\{s_i\}_{i\geq0} \notin \Omega(h)$ for any proper divisor h of f. We shall show that f is unique and satisfies (3.18). Since the period of $\{s_i\}_{i\geq0}$ is p, Corollary 3.8 implies that for some $\tau(x)$ with degree $(\tau) <$ degree (f),

$$\frac{S^{(p)}(x)}{1+x^p} = S(x) = \frac{\tau(x)}{f^*(x)}\,.$$

By our assumption on f and by Lemma 3.21 $gcd\,(f^*(x),\tau(x)) = 1$, so

$$gcd\,\left(f^*(x), \frac{S^{(p)}(x)\cdot f^*(x)}{1+x^p}\right) = 1\,.$$

It follows that

$$gcd\,(f^*(x)(1+x^p), S^{(p)}(x)f^*(x)) = 1 + x^p$$

i.e.

$$f^*(x) \, gcd \, (1+x^p, S^{(p)}(x)) = 1 + x^p \ .$$

Hence

$$f^*(x) = \frac{1 + x^p}{gcd \, (S^{(p)}(x), 1+x^p)} \ . \qquad \qquad \Box$$

Corollary 3.23 The linear equivalence of a binary, periodic sequence $\{s_i\}_{i\geq 0}$ with period p is equal to p − degree $(gcd \, (S^{(p)}(x), 1 + x^p))$, where $S^{(p)}(x) = s_0 + s_1 x + \cdots + s_{p-1} x^{p-1}$.

Example 3.24 Let $\{s_i\}_{i\geq 0}$ have period 15 and let $S^{(15)}(x) = 1 + x^4 + x^7 + x^8 + x^{10} + x^{12} + x^{13} + x^{14}$. Then

$$gcd \, (x^{15}+1, S^{(15)}(x)) = (1+x)(1+x+x^2)(1+x+x^2+x^3+x^4)(1+x+x^4)$$

So $m^*(x) = 1 + x^3 + x^4$, and thus $m(x) = 1 + x + x^4$. Indeed $S(x)$ is the output sequence of the LFSR in Example 3.2.

Corollary 3.23 helps the designer of a non-linear system to determine how safe his system is against the attack descibed by (3.17). The cryptanalist on the other hand who knows part of the sequence, say s_i, $0 \leq i < k$, can try to make the smallest LFSR that generates $s_0, s_1, \ldots, s_{k-1}$ and "predict" s_k with this LFSR.

Definition 3.25 $L_k(\{s_i\}_{i\geq 0})$ is the length of the shortest LFSR that generates $s_0, s_1, \ldots, s_{k-1}$. When it is clear from the context which $\{s_i\}_{i\geq 0}$ is involved we shall simply write L_k. The characteristic polynomial of an L_k-stage LFSR that generates s_i, $0 \leq i \leq k - 1$, will be denoted by $c^{(k)}(x)$.

Clearly $L_k(\{s_i\}_{i\geq 0}) \leq k$ for any sequence $\{s_i\}_{i\geq 0}$, since $(s_0, s_1, \ldots, s_{k-1})$ taken as starting state of any k-state LFSR, will generate $s_0, s_1, \ldots, s_{k-1}$.

Lemma 3.26 Let $\{s_i\}_{i\geq 0}$ be such that $s_0 = s_1 = \cdots = s_{k-2} = 0$ and $s_{k-1} = 1$. Then $L_k(\{s_i\}_{i\geq 0}) = k$.

Proof: An LFSR of length $n < k$ will only output n consecutive zeros if the initial state is the allzero vector. But then the whole output sequence will consist of zeros.

Lemma 3.27 $L_k(\{s_i + t_i\}_{i\geq 0}) \leq L_k(\{s_i\}_{i\geq 0}) + L_k(\{t_i\}_{i\geq 0})$.

Proof: This is a direct consequence of Lemma 3.9. $\qquad \qquad \Box$

It follows from Definition 3.25 that $L_{k+1} \geq L_k$ for any sequence $\{s_i\}_{i\geq 0}$. But more can be said.

Lemma 3.28 If $c^{(k+1)}(x) \neq c^{(k)}(x)$, then $L_{k+1} \geq \max\{L_k, k+1-L_k\}$.

Proof: We already know that $L_{k+1} \geq L_k$. Let $\{t_i\}_{i\geq 0}$ be a sequence starting with k zeros followed by a one. Since $L_{k+1} \geq L_k$, one knows that $L_{k+1}(\{s_i + t_i\}_{i\geq 0}) = L_k(\{s_i\}_{i\geq 0}) = L_k$. The statement now follows from Lemma 3.27 and $k + 1 = L_{k+1}(\{t_i\}_{i\geq 0}) \leq L_{k+1}(\{s_i\}_{i\geq 0}) + L_{k+1}(\{s_i + t_i\}_{i\geq 0}) = L_{k+1} + L_k$.   □

The following theorem shows that in fact equality holds in Lemma 3.28. The proof follows from the Berlekamp-Massey algorithm, that constructs $c^{(k)}(x)$ recursively [Mas69]. This algorithm is well-known in algebraic coding theory for the decoding of BCH codes (see [Ber68], §7).

Theorem 3.29 If $c^{(k+1)}(x) \neq c^{(k)}(x)$, then

$$L_{k+1} = \max \{L_k, k + 1 - L_k\}.$$

Berlekamp-Massey algorithm
Define $L_0 = 0$ and $c^{(0)}(x) = 1$.
The sequence $0, 0, \ldots, 0$ of length k can be generated by the degenerate LFSR with characteristic polynomial $c^{(k)}(x) = 1$ of degree $L_k = 0$. An output sequence $0, 0, \ldots, 0, 1$ of length $k + 1$ can be generated by any $(k + 1)$-stage LFSR, but not by a shorter LFSR as we already saw in Lemma 3.26. So in this case $L_{k+1} = k + 1 = k + 1 - L_k$. This proves the first induction step. We now proceed by induction. So we assume that for $L_k \leq j \leq k - 1$,

$$\sum_{i=0}^{L_k - 1} c_i^{(k)} s_{j - L_k + i} = s_j. \tag{3.19}$$

If (3.19) also holds for $j = k$, then $L_{k+1} = L_k$, $c^{(k+1)}(x) = c^{(k)}(x)$ and there remains nothing to prove. If

$$\sum_{i=0}^{L_k - 1} c_i^{(k)} s_{k - L_k + i} = s_k + 1, \tag{3.20}$$

then by Lemma 3.28 $L_{k+1} \geq \max \{L_k, k + 1 - L_k\}$. We shall prove that equality holds. Let m be the unique integer smaller than k defined by

i) $L_m < L_k$,

ii) $L_{m+1} = L_k$. $\tag{3.21}$

Because we already proved the first induction step, this number m is well defined. It follows that

$$\sum_{i=0}^{L_m - 1} c_i^{(m)} s_{j - L_m + i} = \begin{cases} s_j & \text{if} & L_m \leq j \leq m - 1, \\ s_m + 1 & \text{if} & j = m. \end{cases} \tag{3.22}$$

Notice that $L_k = L_{m+1} = \max \{L_m, m + 1 - L_m\} = m + 1 - L_m$. Define $L = \max \{L_k, k + 1 - L_k\}$. We claim that

$$c(x) = x^{L-L_k} c^{(k)}(x) + x^{L-(k+1-L_k)} c^{(m)}(x) = x^{L-L_k} c^{(k)}(x) + x^{L-k+m-L_m} c^{(m)}(x) \qquad (3.23)$$

will be a suitable choice for $c^{(k+1)}(x)$. Clearly the first term in (3.23) has degree $(L - L_k) + L_k = L$ and the second term has degree $(L - k + m - L_m) + L_m = L - k + m < L$. So $c(x)$ has the right degree. But also by (3.19)-(3.22)

$$\sum_{i=0}^{L-1} c_i \, s_{j-L+i} = \sum_{i=L-L_k}^{L-1} c_{i-(L-L_k)}^{(k)} \, s_{j-L+i} + \sum_{i=L-k+m-L_m}^{L-k+m} c_{i-(L-k+m-L_m)}^{(m)} \, s_{j-L+i} =$$

$$= \sum_{i=0}^{L_k-1} c_i^{(k)} \, s_{j-L_k+i} + \sum_{i=0}^{L_m-1} c_i^{(m)} \, s_{j-L_m-k+m+i} + s_{j-k+m} =$$

$$= \begin{cases} s_j + \quad 0 \quad = s_j, & L \le j \le k - 1, \\ (s_k+1) + 1 \quad = s_k, & j = k. \end{cases}$$

$\square$

Theorem 3.29 only proves that the degree of $c^{(k)}(x)$ is unique. In general the polynomial $c^{(k)}(x)$ will not be unique.

Example 3.30

k	s_k	L_k	$c^{(k)}(x)$	LFSR with initial state
0	1	0	1	—
1	0	1	$1+x$	
2	1	1	x	
3	0	2	$1+x^2$	
4	0	2	$1+x^2$	
5	—	3	x^2	

The problem with non-linear feedback shift registers in general is the difficulty of their analysis. One has to answer questions like: how many different cycles of output sequences are there, what is their length, what is their linear equivalence, etc. . The following will make it clear that it is possible to say at least a little bit about general non-linear feedback shift registers. Clearly the output sequence of a non-linear FSR does not have maximal period if there are two different states with the same successor state. The latter is called a *branch point*.

Theorem 3.31 An n-stage feedback shift register with (non-linear) feedback function $f(s_0, s_1, \ldots, s_{n-1})$ has no branch points iff $f(s_0, s_1, \ldots, s_{n-1}) = s_0 + g(s_1, s_2, \ldots, s_{n-1})$ for some boolean function g in $n-1$ variables.

Proof:

Since f is a Boolean function one can write $f(s_0, s_1, \ldots, s_{n-1}) = g(s_1, s_2, \ldots, s_{n-1}) + s_0 h(s_1, s_2, \ldots, s_{n-1})$.

$\Rightarrow$ If $h(s_1, s_2, \ldots, s_{n-1}) = 0$ for some $(s_1, s_2, \ldots, s_{n-1})$ then $(0, s_1, s_2, \ldots, s_{n-1})$ and $(1, s_1, s_2, \ldots, s_{n-1})$ have the same successor states. Thus a branch point would exist. We conclude that $h \equiv 1$.

$\Leftarrow$ The state $(0, s_1, s_2, \ldots, s_{n-1})$ has successor $(s_1, s_2, \ldots, s_n)$ with $s_n = g(s_1, s_2, \ldots, s_{n-1})$, while $(1, s_1, s_2, \ldots, s_{n-1})$ has successor $(s_1, s_2, \ldots, s_{n-1}, 1 + s_n)$. So there are no branch points. □

Instead of using one non-linear shift register, one can also combine several LFSR's in a non-linear way. Unfortunately not too much is published about these techniques in the open literature. A first idea could be to multiply the output of two LFSR's. However in this way roughly 1/4 of the output bits will be one, so C1 is violated. Adding the output sequences of two LFSR's is not a good idea either, in view of Lemma 3.9. We shall now briefly discuss a method proposed in [Jen83].

Multiplexing

In Figure 3.4 one sees the general picture. The values of a subset A of the inputs determine an address. This address selects one of the remaining inputs (subset B) and gives its value as output.

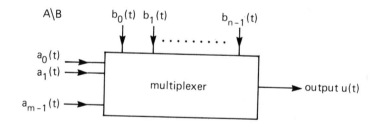

Figure 3.4 Multiplexer

In [Jen83] the following multiplexer is proposed.

1) Take two LFSR's with m resp. n stages, both with a primitive characteristic polynomial and with a non-zero starting state.

2) Choose h distinct elements from $\{0, 1, \ldots, m-1\}$ and order them, say $0 \le i_1 < i_2 < \cdots < i_h < m$. A restriction on the size of h follows in 3). The h-tuple $(a_{i_1}(t), a_{i_2}(t), \ldots, a_{i_h}(t))$ of inputs at time t defines the value

$$N(t) = \sum_{j=1}^{h} a_{i_j}(t) \cdot 2^{j-1}.$$

When $h < m$, $N(t)$ can have any value in $\{0, 1, \ldots, 2^h - 1\}$. When $h = m$ the value of $N(t)$ lies in $\{1, 2, \ldots, 2^h - 1\}$.

3) Let τ be any one-to-one mapping from $\{0, 1, \ldots, 2^h - 1\}$ into $\{0, 1, \ldots, n - 1\}$, if $h < m$ and from $\{1, 2, \ldots, 2^h - 1\}$ into $\{0, 1, \ldots, n - 1\}$ if $h = m$. To be able to do this, we have to add the restriction $2^h \le n$, if $h < m$, resp. the restriction $2^h - 1 \le n$, if $h = m$.

4) The output $u(t)$ at time t is now defined by

$$u(t) = b_{\tau(N(t))}(t). \tag{3.24}$$

Example 3.32: $m = 3, n = 4, h = 2, i_1 = 0, i_2 = 1.$

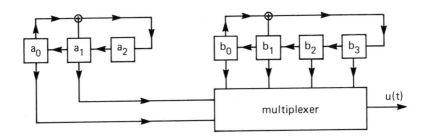

Let τ be defined by $\tau(0) = 3, \tau(1) = 2, \tau(2) = 0, \tau(3) = 1$. With starting states $(1,0,0)$ resp. $(1,0,0,0)$ we get the following sequence:

t	m-register	$N(t)$	$\tau(N(t))$	n-register	$u(t)$
0	100	1	2	1000	0
1	001	0	3	0001	1
2	010	2	0	0010	0
3	101	1	2	0100	0
4	011	2	0	1001	1
5	111	3	1	0011	0
6	110	3	1	0110	1
7	100	1	2	1101	0
8	001	0	3	1010	0
9	010	2	0	0101	0
.	. . .	.	.	. . . .	.

In [Jen83] one can find the following theorems about these multiplexed sequences. They demonstrate how far one can go with their analysis.

Theorem 3.33 If $(m,n) = 1$ the multiplexed sequence has period $(2^m - 1)(2^n - 1)$.

Theorem 3.34 If $(m,n) = 1$ the multiplexed sequence has linear equivalence:

$$
\begin{cases}
= n(1 + m), & \text{if } h = 1, \\[2mm]
= n(1 + m + \binom{m}{2}), & \text{if } h = 2 < m, \\[2mm]
\leq n \sum_{i=0}^{h} \binom{m}{i}, & \text{if } 2 < h < m - 1, \text{with equality} \\
& \text{if the } h \text{ stages are equally spaced,} \\[2mm]
= n(2^m - 1) & \text{if } h = m - 1 \text{ or } h = m.
\end{cases}
$$

For results about properties C1-C3 we refer the reader to [Bek82, § 6.4.4].

Problems

1.   Let $\{s_i\}_{i \geq 0}$ be a binary, periodic sequence of period 8. Let the first eight symbols of $\{s_i\}_{i \geq 0}$ be given by 01001101.

   To which extent does $\{s_i\}_{i \geq 0}$ satisfy Golomb's Randomness Postulates? Apart from the short period and the partial failure in satisfying the Golomb's Randomness Postulates, why else is $\{s_i\}_{i \geq 0}$ not a good pseudo random sequence?

2.   Consider the binary polynomials $f(x) = x^3 + x + 1$ and $g(x) = x^3 + x^2 + 1$. Let $\Omega(f)$, $\Omega(g)$ and $\Omega(fg)$ denote the sets of output sequences of the LFSR's with feedback functions f resp. g and fg. Prove that

$$\Omega(fg) = \{ \ \{u_i + v_i\}_{i \geq 0} \mid \{u_i\}_{i \geq 0} \in \Omega(f) \ \wedge \ \{v_i\}_{i \geq 0} \in \Omega(g) \ \}$$

3.   Let $\{u_i\}_{i \geq 0}$ and $\{v_i\}_{i \geq 0}$ be the output sequences of binary LFSR's of length m resp. n, where $m \geq 2$ and $n \geq 2$. Assume that $\{u_i\}_{i \geq 0}$ and $\{v_i\}_{i \geq 0}$ are both PN sequences and that $gcd(m,n) = 1$. Hence also $gcd(2^m - 1, 2^n - 1) = 1$. Let the sequence $\{w_i\}_{i \geq 0}$ be defined by $w_i = u_i \, v_i$, $i \geq 0$, and let p be the period of $\{w_i\}_{i \geq 0}$.

   a)   Prove that p is a divisor of $(2^m - 1)(2^n - 1)$.

   b)   How many zeros and how many ones appear in a subsequence of length $(2^n - 1)(2^m - 1)$ in the sequence $\{w_i\}_{i \geq 0}$?

c) Prove that $(2^m - 1)(2^n - 1) / p$ must divide the two numbers, determined in b).

d) Prove that $p = (2^m - 1)(2^n - 1)$.

e) How many gaps of length 1 does the $\{w_i\}_{i \geq 0}$-sequence have per period when $m \geq 4$ and $n \geq 4$?

4. Consider the following binary shift register.

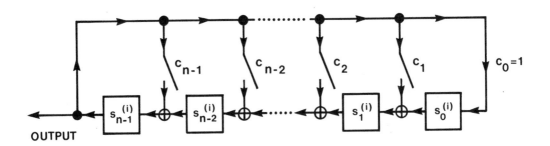

Let $\underline{s}^{(i)} = (s_{n-1}^{(i)}, s_{n-2}^{(i)}, \cdots, s_1^{(i)}, s_0^{(i)})^T$ be the state of the shift register at time i, $i \geq 0$.

a) Give the $n \times n$ matrix T for which $\underline{s}^{(i+1)} = T \underline{s}^{(i)}$ for all $i \geq 0$.

b) Prove that the characteristic equation of T over $\mathbb{R}$ is given by

$$\lambda^n = c_{n-1} \lambda^{n-1} + c_{n-2} \lambda^{n-2} + \cdots + c_1 \lambda + c_0 .$$

From matrixtheorie we may conclude that over $\mathbb{R}$

$$T^n = c_{n-1} T^{n-1} + c_{n-2} T^{n-2} + \cdots + c_1 T + c_0 I , \qquad (*)$$

where I is the $n \times n$ indentity matrix. Since all elements in $(*)$ are integer, equation $(*)$ also holds modulo 2.

c) Derive a recurrence relation between $\underline{s}^{(i+n)}, \underline{s}^{(i+n-1)}, \cdots, \underline{s}^{(i+1)}$ and $\underline{s}^{(i)}$.

d) Which LFSR of length n gives the same output sequence as the above shift register. What does the initial state have to be in this LFSR to generate the same output sequence.

5. Let $\alpha \in GF(2^3)$ be a zero of $f(x) = x^3 + x + 1$. So (by Theorem B.38)
$f(x) = (x - \alpha)(x - \alpha^2)(x - \alpha^4)$ and $f^*(x) = (x - \alpha^3)(x - \alpha^5)(x - \alpha^6) =$
$= (1 - \alpha x)(1 - \alpha^2 x)(1 - \alpha^4 x)$.
Prove that

$$\Omega(f) = \left\{ \sum_{i=0}^{\infty} (u \alpha^i + u^2 \alpha^{2i} + u^4 \alpha^{4i}) x^i \ \middle| \ u \in GF(2^3) \right\}.$$

4 SHANNON THEORY

In Chapter II we have seen that the cryptanalysis of a cryptosystem often depends on the structure that is present in most texts. For instance in Table 2.1 we could find the key 19, because BRUTUS was the only word in the table that made sense. This structure in the plaintext remains present in the ciphertext (it is hoped in a hidden form). If the extra information arising from this structure exceeds our uncertainty about the key, one may be able to determine the ciphertext!

We shall first quantify the concept of information. Let X be a random variable defined on a set $\{x_1, x_2, \ldots, x_n\}$ by $Pr_X\{X = x_i\} = p_i, 1 \le i \le n$. We claim that

$$J(p_i) = -\log_2 Pr_X\{X = x_i\} = -\log_2 p_i \qquad (4.1)$$

is a measure for the amount of *information*, given by the occurrence of the event x_i, $1 \le i \le n$. The base 2 in (4.1) is completely arbitrary. With this base 2 the unit of information is a *bit*. For some values of $Pr_X\{X = x\}$ our claim is intuitively clear. Indeed the occurrence of an event x, that occurs with probability 1, gives no information. So $J(1) = 0$. An event that occurs with probability 1/2, like the specific sex of a newborn baby (assuming that both sexes have the same probability 1/2), gives 1 bit of information. For instance 1 denotes a boy and 0 denotes a girl. So $J(1/2) = 1$. If an event occurs with probability 1/4, then its occurence gives 2 bits of information. This is clear in the case that there are 4 possible outcomes, each with probability 1/4. Each outcome can be represented by a different sequence of two bits. On the other hand the amount of information that an event, which has probability 1/4 to occur, gives, should be independent of the probabilities of the other possible outcomes. So $J(1/4) = 2$. Continuing in this way one gets

$$J(1/2^k) = k, \quad k \geq 0. \tag{4.2}$$

The expected value of $J(Pr_X\{X = x\})$ is called the *entropy* of X and will be denoted by $H(X)$ or $H(\underline{p})$, $\underline{p} = (p_1, p_2, \ldots, p_n)$.

$$H(\underline{p}) = H(X) = E(J(Pr_X\{X = x\})) = \sum_{i=1}^{n} p_i J(p_i) = -\sum_{i=1}^{n} p_i \log_2 p_i. \tag{4.3}$$

One can give the following interpretations to $H(X)$:
- the expected amount of information from a realization of X,
- our uncertainty about X,
- the expected number of bits needed to descibe an outcome of X.

With this interpretation in mind one expects the entropy function $H(X)$ to have the following properties:

P1: $\quad H(p_1, p_2, \ldots, p_n) = H(p_1, p_2, \ldots, p_n, 0)$,

P2: $\quad H(p_1, p_2, \ldots, p_n) = H(p_{\sigma(1)}, p_{\sigma(2)}, \ldots, p_{\sigma(n)})$, for any permutation σ of $\{1, 2, \ldots, n\}$,

P3: $\quad 0 \leq H(p_1, p_2, \ldots, p_n) \leq H(1/n, 1/n, \ldots, 1/n)$,

P4: $\quad H(1/2, 1/2) < H(1/3, 1/3, 1/3) < H(1/4, 1/4, 1/4, 1/4) < \cdots$,

P5: $\quad H(p_1, p_2, \ldots, p_n) = H(p_1, p_2, \ldots, p_{n-2}, p_{n-1} + p_n) +$

$$(p_{n-1} + p_n) H\left(\frac{p_{n-1}}{p_{n-1} + p_n}, \frac{p_n}{p_{n-1} + p_n}\right).$$

Although we shall not prove it here, it can be shown [Khi57] that (4.1) is the only continuous function satisfying (4.2) that yields an entropy function with properties P1-P5. If $n = 2$ one usually writes $h(p)$ instead of $H(p, 1 - p)$. It follows that

$$h(p) = -p \cdot \log_2 p - (1 - p) \cdot \log_2(1 - p), \quad 0 \leq p \leq 1, \tag{4.4}$$

where by definition $h(0) = h(1) = 0$ to make $h(p)$ continuous on the closed interval [0,1].

Example 4.1 Consider the flipping of a coin. Let $Pr\{heads\} = p$ and $Pr\{tails\} = 1 - p$, $0 \leq p \leq 1$. Then the entropy is given by (4.4). That $h(1/2) = 1$ is of course confirmed by the fact that one needs one bit to represent the outcome of the tossing of a fair coin. For instance $0 = heads$ and $1 = tails$. Since $h(1/4) \approx 0.8113$ one expects that on the average only 0.8113 bits are needed to represent the outcome of a tossing of an unfair coin with $Pr\{heads\} = 1/4$. This is true in the sense that one can approach the number 0.8113 arbitrarily closeby. In Chapter V we shall show how to do this. The trick will be to represent the outcome of n consecutive tossings by a string of bits. For instance with $n = 2$ one can represent the outcomes as follows:

2 succ. outcomes	probability	representation
h h	1/16	111
h t	3/16	110
t h	3/16	10
t t	9/16	0

The expected length of a representation is

$$3 \cdot \frac{1}{16} + 3 \cdot \frac{3}{16} + 2 \cdot \frac{3}{16} + 1 \cdot \frac{9}{16} = \frac{27}{16} \, .$$

But each representation describes two outcomes, so this scheme needs $27/32 \approx 0.843$ bits per tossing. Taking $n = 3, n = 4$, etc. gives better approximations of $h(1/4)$. There is however the problem that the receiver of a long string of zeros and ones should be able to determine the outcomes of the tossings in a unique way. One can easily verify that any sequence made up from the subsequences 111, 110, 10 and 0, can only in one way be broken up into these subsequences. We come back to this problem in Chapter V.

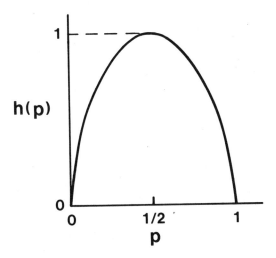

Figure 4.1 The function $h(p)$, $0 \le p \le 1$.

Example 4.2 The 26 letters in the English alphabet can be represented with $\log_2 26 \approx 4.70$ bits per letter, by coding sufficiently long strings of letters into binary strings. On the other hand, the entropy of 1-grams can easily be computed with the probabilities given in Table 1.2. One obtains 4.15 bits per letter. Also for $n = 2$ and 3 these computations have been made (see [Mey82, app.F]). One gets the following values:

H (1-grams) $\approx$ 4.15 bits/letter

H (2-grams)/2 $\approx$ 3.62 bits/letter

H (3-grams)/3 $\approx$ 3.22 bits/letter

According to some tests the asymptotical value for $n \to \infty$ is 1.5 bits/letter!

Definition 4.3 Let $(X_0, X_1, \ldots, X_{n-1})$, $n \geq 1$, denote the plaintext generated by a plaintext source S over the alphabet $\mathbb{Z}_q$. Then the *redundancy* D of $(X_0, X_1, \ldots, X_{n-1})$ is defined by

$$D_n = n \cdot \log_2 q - H(X_0, X_1, \ldots, X_{n-1}). \tag{4.5}$$

The quantity $\delta_n = D/n$ is called the *average redundancy* per plaintext letter.

The redundancy in a text measures how much the length of the text (when translated into a binary string) exceeds the length that is strictly necessary to carry the information of the text.

Definition 4.4 In a ciphertext-only attack on a cryptographic system with key-space K and plaintext source S the *unicity distance of plaintext* is defined by

$$\min \{ n \in N \mid D_n \geq H(K) \}. \tag{4.6}$$

As soon as the redundancy in the plaintext exceeds the uncertainty about the key, the cryptanalist with sufficient resources may be able to determine that plaintext from the ciphertext. So the unicity distance tells the user of a cryptosystem when to change the key in order to keep the system safe.

Example 4.2 (continued). With a simple substitution in an English text, one has $H(K) = \log_2 26! \approx 88.382$ bits, assuming that all 26! possible substitutions are equally likely. If one approximates the redundancy D_n in n letters by $(4.70 - 1.50)n = 3.20n$ bits, one obtains a unicity distance of 28. According to Friedman [Fri73]: "practically every example of 25 or more characters representing the monoalphabetic substitution of a 'sensible' message in English can be readily solved." These two numbers are in remarkable agreement.

Let X and Y be two random variables. The joint distribution of X and Y is denoted by

$$Pr_{X,Y} \{ X = x, Y = y \} = p_{X,Y}(x, y). \tag{4.7}$$

Similarly the conditional probability that $X = x$, given $Y = y$, is denoted by

$$Pr_{X,Y} \{ X = x \mid Y = y \} = p_{X \mid Y}(x \mid y). \tag{4.8}$$

The uncertainty about X given $Y = y$ is defined analogous to (4.3) by

$$H(X \mid Y = y) = -\sum_x p_{X \mid Y}(x \mid y) \cdot \log_2 p_{X \mid Y}(x \mid y). \tag{4.9}$$

It can be interpreted as the expected amount of information that a realization of X gives, when the occurrence of $Y = y$ is already known. The *equivocation* $H(X \mid Y)$ or *conditional entropy* of X given Y is the expected value of $H(X \mid Y = y)$. So

$$H(X \mid Y) = \sum_y p_Y(y) \cdot H(X \mid Y = y) =$$

$$= -\sum_y p_Y(y) \cdot \sum_x p_{X \mid Y}(x \mid y) \cdot \log_2 p_{X \mid Y}(x \mid y) =$$

$$= -\sum_x \sum_y p_{X,Y}(x,y) \cdot \log_2 p_{X \mid Y}(x \mid y). \tag{4.10}$$

Theorem 4.5 $H(X,Y) = H(Y) + H(X \mid Y) = H(X) + H(Y \mid X)$.

Proof: We use $p_{X,Y}(x,y) = p_Y(y) \cdot p_{X \mid Y}(x \mid y)$, (4.10) and the analog of (4.3) for two variables.

$$H(X,Y) = -\sum_x \sum_y p_{X,Y}(x,y) \cdot \log_2 p_{X,Y}(x,y) =$$

$$-\sum_x \sum_y p_{X,Y}(x,y) \cdot \log_2 p_Y(y) - \sum_x \sum_y p_{X,Y}(x,y) \cdot \log_2 p_{X \mid Y}(x \mid y) =$$

$$-\sum_y p_Y(y) \cdot \log_2 p_Y(y) + H(X \mid Y) = H(Y) + H(X \mid Y).$$

The second equality follows by symmetry. ▯

In words, Theorem 4.5 states that the expected number of bits needed to describe a realization of X and Y equals the expected number of bits needed to descibe a realization X plus the expected number of bits needed to descibe Y when X is already known.

Corollary 4.6 If X and Y are independent then

i) $H(X,Y) = H(X) + H(Y)$,

ii) $H(X \mid Y) = H(X)$,

iii) $H(Y \mid X) = H(Y)$.

Proof: Repeat the proof of Theorem 4.5 with $p_{X,Y}(x,y) = p_X(x) \cdot p_Y(y)$. ▯

The amount of information $I_{X;Y}(x;y)$ that the realization $Y = y$ gives about the realization $X = x$ is the amount of information that the occurrence of x gives minus the amount of information that $X = x$ gives when $Y = y$ is already known. I.e.

$$I_{X;Y}(x;y) = -\log_2 p_X(x) - \{-\log_2 p_{X|Y}(x \mid y)\} =$$

$$-\log_2 \frac{p_X(x)}{p_{X|Y}(x \mid y)} = -\log_2 \frac{p_X(x) \cdot p_Y(y)}{p_{X,Y}(x,y)} = I_{Y;X}(y;x). \tag{4.11}$$

The *mutual information* $I(X;Y)$ of X and Y is the expected value of $I_{X;Y}(X;Y)$, i.e.

$$I(X;Y) = \sum_x \sum_y p_{X,Y}(x,y) \cdot I_{X;Y}(x;y) =$$

$$= -\sum_x \sum_y p_{X,Y}(x,y) \cdot \log_2 \frac{p_X(x) \cdot p_Y(y)}{p_{X,Y}(x,y)} = \tag{4.12}$$

$$= -\sum_x \sum_y p_{X,Y}(x,y) \cdot \log_2 \frac{p_X(x)}{p_{X|Y}(x \mid y)}.$$

Theorem 4.7

$$I(X;Y) = H(X) + H(Y) - H(X,Y) = H(X) - H(X \mid Y) = H(Y) - H(Y \mid X).$$

Proof: From (4.12) it follows that

$$I(X;Y) = -\sum_x \sum_y p_{X,Y}(x,y) \cdot \log_2 \frac{p_X(x)}{p_{X|Y}(x \mid y)} =$$

$$-\sum_x \sum_y p_{X,Y}(x,y) \cdot \log_2 p_X(x) + \sum_x \sum_y p_{X,Y}(x,y) \cdot \log_2 p_{X|Y}(x \mid y) =$$

$$-\sum_x p_X(x) \cdot \log_2 p_X(x) - H(X \mid Y) = H(X) - H(X \mid Y).$$

The other statements follow from Theorem 4.5.   ▯

$I(X;Y)$ can be interpreted as the expected amount of information that Y gives about X (or X about Y).

Example 4.8 The *binary symmetric channel BSC* can be described as follows. A source sends $X = 0$ or $X = 1$, each with probability 1/2. The receiver gets $Y = X$ with probability $1 - p$ and $Y = 1 - X$ with probability p. It follows that

$$p_Y(0) = p_{Y|X}(0 \mid 0) \cdot p_X(0) + p_{Y|X}(0 \mid 1) \cdot p_X(1) = (1-p)/2 + p/2 = 1/2.$$

Similarly $p_Y(1) = 1/2$. Also $p_{X,Y}(0,0) = p_{X,Y}(1,1) = (1-p)/2$ and $p_{X,Y}(0,1) = p_{X,Y}(1,0) = p/2$. So for the *BSC*

$$I(X;Y) = -2 \cdot \{\frac{1-p}{2} \log_2 \frac{1/2}{1-p} + \frac{p}{2} \log_2 \frac{1/2}{p}\} =$$

$$1 + p \cdot \log_2 p + (1-p) \cdot \log_2 (1-p) = 1 - H(p).$$

So the receiver gets $1 - H(p)$ bits of information about X per received symbol Y. How to approach this quantity $1 - H(p)$ is the fundamental problem in algebraic coding theory [Mac77, Ch.1. §6]. For $p = 1/2$ the receiver gets no information about the transmitted symbols, as is to be expected.

Let us now return to the conventional cryptosystem as explained in Chapter I. Assume that a probability distribution $Pr_K(K = k)$ is defined on the keyspace and let

$M^n = (M_0, M_1, \ldots, M_{n-1})$ denote the plaintext, and

$C^v = (C_0, C_1, \ldots, C_{v-1})$ denote the ciphertext.

So $C^v = E_K(M^n)$. Since E_K is a one-to-one mapping, one has

$$H(M^n \mid K, C^v) = 0. \tag{4.13}$$

Of course the user of the cryptosystem is interested in how much information C^v contains about M^n.

Theorem 4.9 $I(M^n; C^v) \geq H(M^n) - H(K)$.

Proof: By (4.13) and Theorem 4.5 one has that

$$H(K \mid C^v) = H(K \mid C^v) + H(M^n \mid K, C^v) = H(M^n, K \mid C^v) =$$

$$H(M^n \mid C^v) + H(K \mid M^n, C^v) \geq H(M^n \mid C^v).$$

In words: given the ciphertext the uncertainty about the key is at least as great as the uncertainty about the plaintext. It follows that

$$H(M^n \mid C^v) \leq H(K \mid C^v) \leq H(K)$$

and by Theorem 4.7 that

$$I(M^n; C^v) = H(M^n) - H(M^n \mid C^v) \geq H(M^n) - H(K).$$

□

Definition 4.10 A cryptosystem is called *unconditionally secure* or is said to have *perfect secrecy* if $I(M^n; C^v) = 0$.

Corollary 4.11 A necesarry condition for the perfect secrecy of a cryptosystem is given by $H(M^n) \leq H(K)$.

Example 4.12 Suppose that we have 2^k keys, all with probability $1/2^k$. Then

$$H(K) = -\sum_{i=1}^{2^k} 2^{-k} \cdot \log_2 2^{-k} = k \text{ bits.}$$

If the messages are the outcome of n tossings with a fair coin, one has in a similar way that $H(M^n) = n$. So for perfect secrecy one needs $k \geq n$.

The reader is invited to give a formal proof of the perfect secrecy of the Vernam cipher. As the title of this chapter suggests most of the material that has been discussed above is due to C.E.Shannon [Shn49]. He also describes a "random cryptosystem", with an equivocation of key $H(K \mid C)$ that can be lowerbounded by a certain quantity. We omit that discussion here. The statements below Definition 4.4 can be formally proved for random cryptosystems. The interested reader is also referred to [Hel77], where it is shown that cryptosystems do exist that are better than a "random" cryptosystem.

Problems

1. Assuming that the English language has an information rate of 1.5 bits per letter, what is the unicity distance of the Caesar cipher, when applied to an English text? Answer the same question for the Vigenère cryptosystem with key-length r.

2. Consider a memoryless message source that generates uniformly distributed letters X from $\{0, 1, 2\}$.
    After transmission over a channel the symbol Y, that is received, will be equal to X with probability $1-p$, $0 \leq p \leq 1$, and will be equal to any of the other two letters in the alphabet with probability $p/2$.
    Compute the mutual information $I(X, Y)$.

3. Proof that the one-time pad is unconditionally secure.

5 HUFFMAN CODES

It is clear from the previous chapter (in particular from (4.5) and (4.6)) that the security of a cryptosystem can be increased by reducing the redundancy in the plaintext. In Example 4.1 such a reduction has been demonstrated. In this chapter we shall describe a general method for what is called *data compression* or *source coding*. Let a (plaintext) source S output independently chosen symbols from the set $\{m_1, m_2, \ldots, m_n\}$, with respective probabilities $p_1, p_2, \ldots, p_n$. A symbol m_i, $1 \le i \le n$, will be encoded into a binary string $\underline{c_i}$ of length l_i. The set $\{\underline{c_1}, \underline{c_2}, \ldots, \underline{c_n}\}$ is called a *code* C for the source S. The idea of data compression is to use such a code that the expected value of the length of the encoded plaintext is minimal. Since the output symbols are independent we have to minimize the expected length per symbol

$$L = \sum_{i=1}^{n} p_i l_i. \tag{5.1}$$

over all possible codes C for the source S. There is however an additional constraint. One has to be able to retrieve the individual messages from the concatenation of the succesive codewords. Not every code has this property. Indeed let $C = \{0, 01, 10\}$. The sequence 010 can be made in two ways: 0 followed by 10 and 01 followed by 0. This ambiguity has to be avoided.

Definition 5.1 A code C is called *uniquely decodable* (shortened to U.D.) if every concatenation of codewords can only in one way be split into individual codewords.

Example 5.2 Let $n = 4$ and $C = \{0, 01, 011, 111\}$. This code C is U.D., as we shall now prove.

Consider a concatenation of codewords. If the left most bit is a 1, the left most codeword is 111. If the left most bit is a 0, the concatenation is either of the form 011 . . . 1 or starts like 011 . . . 10 . . .. Let the number of ones after the initial 0 be k. Depending on $k = 3l$, $3l + 1$, or $k = 3l + 2$ the left most codeword is 0, 01 resp. 011. One can now remove this codeword and apply the same decoding rule to the remaining, shorter concatenation of codewords.

Theorem 5.3: (*McMillan inequality*, [McM56])
A necessary and sufficient condition for the existence of a uniquely decodable code C with codewords of length l_i, $1 \le i \le n$, is

$$\sum_{i=1}^{n} \frac{1}{2^{l_i}} \le 1. \tag{5.2}$$

Proof: We shall only prove that (5.2) is necessary for the existence of a U.D. code with codeword lengths l_i, $1 \le i \le n$. That (5.2) is also sufficient will be proved later in this chapter. Let K be equal to the left hand side of (5.2) and let us assume that the lengths are in non-decreasing order, i.e. $l_1 \le l_2 \le \cdots \le l_n$. Then

$$K^N = (\sum_{i=1}^{n} \frac{1}{2^{l_i}})^N = \sum_{j=Nl_1}^{Nl_n} \frac{N_j}{2^j}, \tag{5.3}$$

where N_j is the number of ways to write j as $l_{i_1} + l_{i_2} + \cdots + l_{i_N}$, or alternatively N_j is the number of ways to make a concatenation of N codewords of total length j. Because C is U.D. no two different choices of N concatenated codewords will give rise to the same string of length j. So $N_j \le 2^j$. Substitution of this inequality in (5.3) implies that for all $N \ge 1$

$$K^N \le \sum_{j=Nl_1}^{Nl_n} 1 = N(l_n - l_1) + 1.$$

Since the left-hand side is an exponential function of N, while the right hand side is a linear function of N, we conclude that $K \le 1$. $\square$

As can be seen in Example 5.2 the received sequence often has to be examined much further than the length of the longest codeword to decode it. This is not very practical.

Definition 5.4: A code C is called a *prefix code* or *instantaneous code* if no codeword is a prefix of another codeword.

The code in Example 5.2 is not a prefix code, since the codeword 0 is a prefix of the codeword 01. The code in Example 4.1 clearly is a prefix code. For the decoding of a prefix code one simply looks for a prefix of the received sequence that is a codeword. Because the code is a prefix code this codeword is unique. Remove it and proceed in the same way. This proves Theorem 5.5. Note that we need to examine at most l_n bits of the sequence to determine the first codeword in the received sequence.

Theorem 5.5 A prefix code is uniquely decodable.

Theorem 5.6: (*Kraft inequality*, [Kra49])
A necessary and sufficient condition for the existence of a prefix code with lengths l_i, $1 \leq i \leq n$, is

$$\sum_{i=1}^{n} \frac{1}{2^{l_i}} \leq 1. \tag{5.4}$$

Proof: A prefix code is U.D. by Theorem 5.5. So it follows from the McMillan inequality (Theorem 5.3) that inequality (5.4) is a necessary condition for a code to be a prefix code. We shall now prove that (5.4) implies the existence of a prefix code with lengths l_i, $1 \leq i \leq n$, and a fortiori of a U.D. code with these lengths. Without loss of generality $l_1 \leq l_2 \leq \cdots \leq l_n$. Because of this ordering and since $\sum_{i=1}^{n-1} 1/2^{l_i} < 1$ we can define the vectors $\underline{c}_i = (c_{i,1}, c_{i,2}, \ldots, c_{i,l_i})$, $1 \leq i \leq n$, by

$$\sum_{j=1}^{i-1} \frac{1}{2^{l_j}} = \frac{c_{i,1}}{2} + \frac{c_{i,2}}{2^2} + \cdots + \frac{c_{i,l_i}}{2^{l_i}}. \tag{5.5}$$

So $\underline{c}_1 = (0, 0, \ldots, 0)$ of length l_1, $\underline{c}_2 = (0, 0, \ldots, 0, 1, 0, \ldots, 0)$ of length l_2, with a 1 on coordinate l_1, etc.. By definition $\underline{c}_i$, $1 \leq i \leq n$, has length l_i. It remains to show that no $\underline{c}_u$ can be the prefix of a codeword $\underline{c}_v$. Suppose the contrary. Clearly $l_u \neq l_v$. So $l_u < l_v$ and thus $u < v$. It also follows that

$$\sum_{j=1}^{v-1} \frac{1}{2^{l_j}} - \sum_{j=1}^{u-1} \frac{1}{2^{l_j}} = \sum_{j=l_u+1}^{l_v} \frac{c_{v,j}}{2^j} \leq \sum_{j=l_u+1}^{l_v} \frac{1}{2^j} < \sum_{j=l_u+1}^{\infty} \frac{1}{2^j} = \frac{1}{2^{l_u}},$$

while on the other hand

$$\sum_{j=1}^{v-1} \frac{1}{2^{l_j}} - \sum_{j=1}^{u-1} \frac{1}{2^{l_j}} = \sum_{j=u}^{v-1} \frac{1}{2^{l_j}} \geq \frac{1}{2^{l_u}}.$$

A contradiction! □

It is quite remarkable that the conditions (5.2) and (5.4) in Theorems 5.3 resp. 5.6 are the same. So the smallest expected value of the length among all U.D. codes is equal to the smallest expected value of the length among all prefix codes! The next two theorems give bounds on the expected value of the length of prefix codes or U.D. codes.

Theorem 5.7 Let C be a U.D. code with codewords $\underline{c}_i$ of length l_i for the messages m_i that occur with probability p_i, $1 \leq i \leq n$. Then the expected value L of the length satisfies

$$L = \sum_{i=1}^{n} p_i l_i \geq H(p). \tag{5.6}$$

Proof: It follows from the well-known inequality $\ln x \leq x - 1$, $x > 0$, and from (5.2) that

$$H(p) - L = -\sum_{i=1}^{n} p_i \cdot \log_2 p_i - \sum_{i=1}^{n} p_i \cdot l_i = \frac{1}{\ln 2} \sum_{i=1}^{n} p_i \cdot \ln \frac{1}{p_i \cdot 2^{l_i}} \leq$$

$$\frac{1}{\ln 2} \sum_{i=1}^{n} p_i \cdot (\frac{1}{p_i \cdot 2^{l_i}} - 1) = \frac{1}{\ln 2} (\sum_{i=1}^{n} \frac{1}{2^{l_i}} - 1) \leq 0. \qquad \square$$

Theorem 5.8 Let S be a source generating symbols m_i with probabilities p_i, $1 \leq i \leq n$. Then a prefix code C exists for this source with expected length $L < H(p) + 1$.

Proof: Without loss of generality $p_1 \geq p_2 \geq \cdots \geq p_n$. Define l_i by $\lceil \log_2 1/p_i \rceil$, $1 \leq i \leq n$. Here $\lceil x \rceil$ denotes the smallest integer greater than or equal to x. Clearly $l_1 \leq l_2 \leq \cdots \leq l_n$ and

$$\sum_{i=1}^{n} 1/2^{l_i} \leq \sum_{i=1}^{n} p_i = 1.$$

For these values of l_i, $1 \leq i \leq n$, construct the code C, described in the proof of Theorem 5.6. It is a prefix code with expected value L of its length satisfying

$$L = \sum_{i=1}^{n} p_i \cdot l_i = \sum_{i=1}^{n} p_i \cdot \lceil \log_2 \frac{1}{p_i} \rceil < \sum_{i=1}^{n} p_i (\log_2 \frac{1}{p_i} + 1) = H(p) + 1.$$

$$\square$$

Corollary 5.9 The minimal expected length of all prefix (or U.D.) codes for a source S with probability distribution p has a value L satisfying

$$H(p) \leq L < H(p) + 1. \tag{5.7}$$

In words, L is at least $H(p)$ but does not need to exceed $H(p)$ by one bit or more. By applying Theorem 5.8 to N-tuples of source symbols, one gets in the same way an expected length $L^{(N)}$ per N-gram, satisfying

$$N \cdot H(p) \leq L^{(N)} < N \cdot H(p) + 1.$$

It follows that

$$H(p) \leq \frac{L^{(N)}}{N} < H(p) + \frac{1}{N}. \tag{5.8}$$

So $\lim_{N \to \infty} L^{(N)}/N = H(p)$. This confirms the third interpretation of the entropy function H, that was given at the beginning of Chapter 4.

We shall now discuss an algorithm that finds a prefix code with minimal expected length L.

Theorem 5.10 Consider the source S with symbols m_i, $1 \le i \le n$, and probabilities $p_1 \ge p_2 \ge \cdots \ge p_n$. Let C be a U.D. code for this source with codewords $\underline{c}_i$ of length l_i, $1 \le i \le n$, and expected value L of the length. Let L be minimal among all U.D. codes for this source S. Then after a suitable permutation of the codewords associated with the messages of the same probability,

P1) $l_1 \le l_2 \le \cdots \le l_n$,

P2) C can be assumed to be a prefix code,

P3) $\sum_{i=1}^{n} 1/2^{l_i} = 1$,

P4) $l_{n-1} = l_n$,

P5)   two of the codewords of length l_n differ only in their last coordinate.

Proof:

P1:   Suppose that $p_u > p_v$ and $l_u > l_v$. Make the code C^* from C by interchanging $\underline{c}_u$ and $\underline{c}_v$. Then the expected length L^* of C^* satisfies

$$L^* = L + p_u(l_v - l_u) + p_v(l_u - l_v) = L + (p_u - p_v)(l_v - l_u) < L,$$

while C^* is obviously a U.D. code. This contradicts the minimality of L.

P2:   If a U.D. code exists with expected length L, then a prefix code with the same expected length L also exists by Theorem 5.3 and 5.6.

P3:   If $\sum_{i=1}^{n} 1/2^{l_i} < 1$, one can decrease l_n by 1 and still satisfy (5.4). By Theorem 5.6 a prefix code with smaller expected length would exist. This contradicts our assumption on C.

P4:   If $l_{n-1} < l_n$ one cannot have equality in P3), because of P1).

P5:   Delete the last coordinate of $\underline{c}_n$ and call the resulting vector $\underline{c}_n^*$. Let C^* be the code $\{\underline{c}_1, \underline{c}_2, \ldots, \underline{c}_{n-1}, \underline{c}_n^*\}$. It follows from P3) that C^* does not satisfy (5.4). So C^* is not a prefix code, while C was. This is only possible if $\underline{c}_n^*$ is a proper prefix of some codeword $\underline{c}_i$, $1 \le i \le n-1$. Since $l_i \le l_n$, this proves that $\underline{c}_i$ and $\underline{c}_n$ only differ in their last coordinate. $\square$

The *Huffman algorithm* [Huf52] consists of two parts:

The *reduction* process. Let S be a source which outputs symbols m_i, $1 \le i \le n$, with respective probabilities $p_1 \ge p_2 \ge \cdots \ge p_n$. Replace the two symbols m_{n-1} and m_n by one new symbol with probability $p_{n-1} + p_n$. In this way one obtains a new source with one output symbol less than S. Repeat this reduction process until a source is obtained with an alphabet of size two.

The *splitting* process. Encode the two symbols in the last alphabet with the codewords 0 and 1. Now take a step back in the reduction process. One of the symbols will be split up in two symbols, which will be encoded by simply adding a 0 resp. a 1 at the end. The other symbol keeps its original encoding. Repeat this splitting up process until the original alphabet is back and all the symbols are encoded. The code obtained in this way is called a *Huffman code*.

Example 5.11: $n = 6$.
reduction process:

m_i	p_i				
m_1	0.3	0.3	0.3	0.4	0.6
m_2	0.2	0.2	0.3	0.3	0.4
m_3	0.2	0.2	0.2	0.3	
m_4	0.1	0.2	0.2		
m_5	0.1	0.1			
m_6	0.1				

splitting process:

m_i	$\underline{c}_i$				
m_1	00	00	00	1	0
m_2	10	10	01	00	1
m_3	11	11	10	01	
m_4	011	010	11		
m_5	0100	011			
m_6	0101				

Theorem 5.12 A Huffman code satisfies the properties P1 - P5 in Theorem 5.10.

Proof: The proof is straightforward with an induction argument.

Theorem 5.13 Let S be a source with symbols m_i, $1 \le i \le n$, with resp. probabilities $p_1 \ge p_2 \ge \cdots \ge p_n$. Let C be a Huffman code for this source with lengths l_i, $1 \le i \le n$, and expected length L. Then L is minimal among all U.D. codes for this source.

Proof: For $n = 2$ the statement is obvious. We proceed by induction on n. Let C be a Huffman code for a source S with n symbols. Let the codewords $\underline{c}_i$ of C have length l_i, $1 \le i \le n$, and let L be the expected length of a codeword in C. Let C^* be a U.D. code for S with codewords $\underline{c}_i^*$ of length l_i^*, $1 \le i \le n$, whose expected value L^* of the length is minimal among all U.D. codes for S. By Theorems 5.10 and 5.12 both codes, C and C^*, satisfy properties P1-P5. Without loss of generality $\underline{c}_{n-1}$ and $\underline{c}_n$ in C differ only in their last coordinate, as do $\underline{c}_{n-1}^*$ and $\underline{c}_n^*$ in C^*. Now apply one step of the reduction process to C and C^*. One obtains a Huffman code D and a prefix code D^*, with expected lengths M resp. M^*. By the induction hypothesis $M \le M^*$. The proof now follows from

$$L - L^* = \{ \sum_{i=1}^{n} l_i \cdot p_i \} - \{ \sum_{i=1}^{n} l^*_i \cdot p_i \} =$$

$$\{ \sum_{i=1}^{n} l_i \cdot p_i - p_{n-1} - p_n \} - \{ \sum_{i=1}^{n} l^*_i \cdot p_i - p_{n-1} - p_n \} =$$

$$\{ \sum_{i=1}^{n-2} l_i \cdot p_i + (l_n - 1)(p_{n-1} + p_n) - \{ \sum_{i=1}^{n-2} l^*_i \cdot p_i + (l^*_n - 1)(p_{n-1} + p_n) \} =$$

$$= M - M^* \le 0.$$

If one wants to compress data from a source with unknown statistics, the Huffman algorithm cannot be applied. For this case we refer the reader to [Ziv77], [Eli86] or [Wil86].

Problems

1. Decode the string 011001111111111100011, which has been made with the code in Example 5.2.

2. Apply the Huffman algorithm to the source S that generates the symbols a, b, c, d, e, f, g and h independently with probabilities 1/2, resp. 1/4, 1/8, 1/16, 1/32, 1/64, 1/128 and 1/128. What is the expected number of bits needed for the encoding of one letter? Compare this with the entropy of the source.

6 DES

In 1974 the National Bureau of Standards (NBS) solicited the American industry to develop a cryptosystem that could be used as a standard in unclassified U.S. Government applications. IBM developed a system called LUCIFER. After being modified and simplified, this system became the *Data Encryption Standard* (DES) in 1977. It has been implemented on a chip, which makes it very suitable for use in large communication systems. The encryption and decryption algorithms of DES have been made public. This has never been done before, although in each textbook one can find the remark that the security of a cryptosystem should not depend on the secrecy of the system. Later we shall give a complete description of DES. DES is a block cipher operating on 64 bits simultaneously (see Figure 6.1). Although the keysize is 64, the effective keysize is 56 bits. The remaining 8 bits are parity checks. The input M and key K result in a ciphertext C, which we shall denote by DES (M,K). For the decryption the same DES chip with the same key can be used, as we shall see later.

A disadvantage of the scheme in Figure 6.1 is of course that the same input will yield the same output, when using the same key. This is often undesirable. An eavesdropper that observes how the receiver responds to a certain message can predict that same response when the same ciphertext is transmitted. He can even initiate this response by sending this ciphertext himself! We shall discuss one of the various ways to avoid this problem.

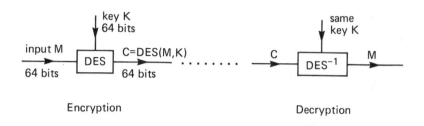

Encryption   Decryption

Figure 6.1 The Data Encryption Standard

Cipherblock chaining:

Divide the plaintext into blocks M_i, $i \geq 1$, of 64 bits. Let C_0 be a initialization vector I. Define $C_{i+1} = \text{DES}(M_{i+1} \oplus C_i, K)$, $i \geq 0$. So the output C_i is added to the input M_{i+1}, componentwise modulo 2, and then fed into the DES-algorithm to generate output C_{i+1}. To decrypt set C_0 equal to the initialization vector I. For all $i \geq 0$ one first computes $\text{DES}^{-1}(C_{i+1}, K) = M_{i+1} \oplus C_i$. To the result one simply adds the previous ciphertext C_i, which is stored in the buffer, in order to retrieve M_{i+1}. If a transmission error occurs, the resulting damage will only be limited to two consecutive blocks of plaintext. Indeed if C_i is not received correctly, then M_i and M_{i+1} will be corrupted. But M_{i+2} will be correctly deciphered, since it only depends on (the correct) C_{i+1} and C_{i+2}.

For other feedback modes we refer the reader to [Bek82], [Den82], [Kon81] or [Mey82].

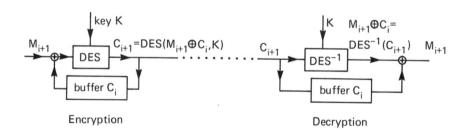

Encryption   Decryption

Figure 6.2 Cipherblock chaining

DES:

The general outline of the DES algorithm is depicted in Figure 6.4. The 64 input bits are first permuted by the permutation *IP* listed in Table 6.3. So the 58-th bit of the input goes to position 1, the 50-th bit goes to position 2, etc.. The left most 32 bits are put in L_0, while the right most 32 bits are put in R_0. After 16 rounds of manipulations the inverse IP^{-1} (see Table 6.3) of the permutation *IP* is applied to generate the final output. We shall now describe what happens during each of the 16 rounds. Let L_{i-1} and R_{i-1}, $1 \le i \le 16$, be the left 32 resp. right 32 bits before round i. Then $L_i = R_{i-1}$ while R_i is defined by

$$R_i = L_{i-1} \oplus f(R_{i-1}, K_i), \tag{6.1}$$

where the $f(R_{i-1}, K_i)$ is a binary vector of length 32 and where the addition is componentwise modulo 2. The vector K_i in (6.1) is a 48-bit vector made from the key K in a way that we shall describe later.

58	50	42	34	26	18	10	2		40	8	48	16	56	24	64	32
60	52	44	36	28	20	12	4		39	7	47	15	55	23	63	31
62	54	46	38	30	22	14	6		38	6	46	14	54	22	62	30
64	56	48	40	32	24	16	8		37	5	45	13	53	21	61	29
57	49	41	33	25	17	9	1		36	4	44	12	52	20	60	28
59	51	43	35	27	19	11	3		35	3	43	11	51	19	59	27
61	53	45	37	29	21	13	5		34	2	42	10	50	18	58	26
63	55	47	39	31	23	15	7		33	1	41	9	49	17	57	25

initial permutation IP   inverse initial permutation IP⁻¹

Table 6.3 *IP* and IP^{-1}

In Figure 6.4 one can see that DES^{-1} can be computed with the same scheme by simply going from the bottom to the top. Indeed $R_{i-1} = L_i$ and L_{i-1} is well defined by

$$L_{i-1} = R_i \oplus f(R_{i-1}, K_i). \tag{6.2}$$

In Figure 6.5 the computation of the function f is depicted. The 32 bits of R_{i-1} are expanded to a 48-bit vector by the bitselection table E, which is listed in Table 6.6. So bit 32 is on position 1 (and position 47), bit 1 on position 2, etc.. To these 48 bits the 48-bit vector K_i is added componentwise. The first 6 bits are fed into the S-box S_1, the next 6 bits are fed into the S-box S_2, etc.. Each S-box outputs 4 bits. The $8 \times 4 = 32$ bits leaving the S-boxes are subsequently permuted according to the permutation P, which is listed in Table 6.6. So the 16-th bit goes to position 1, the 7-th bit to position 2, etc.. The structure of the S-boxes can be seen in Table 6.9. Let $a_1 a_2 \cdots a_6$ be the 6 input bits of S-box S_k, $1 \le k \le 8$. Then the output will be the binary representation of $(S_k)_{ij}$, where i has $a_1 a_6$ as binary representation and j has $a_2 a_3 a_4 a_5$ as binary representation.

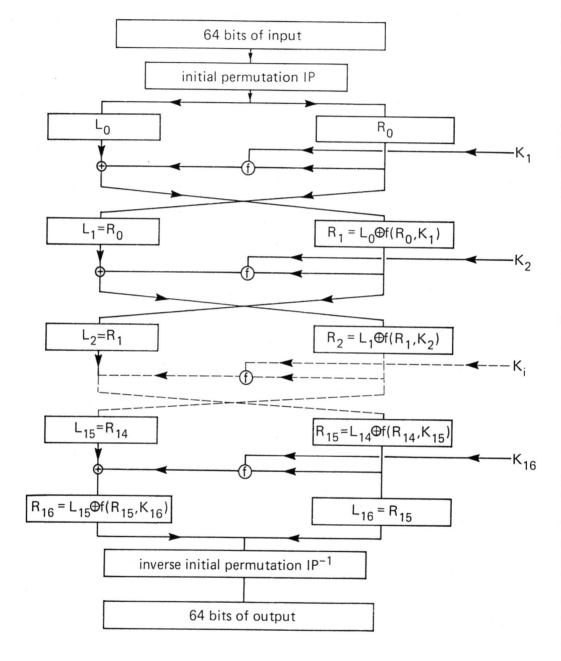

Figure 6.4 DES Encryption algorithm

So the input 000111 in S_1 yields the binary representation of $(S_1)_{1,3} = 4$, thus 0100, as output.

The key K is divided into eight 8-tuples. Each 8-tuple contains an overall parity check bit to detect possible errors that are made during the key generation or storage. To be more precise: bit 8 makes the parity in the first 8 bits odd, bit 16 does the same in the second 8-tuple, etc.. So 1 error per 8-tuple will be detected. These 8 redundant bits are not used for the generation of the vectors K_i, $1 \le i \le 16$. In Figure 6.10 one can see how the 48-bit vectors K_i, $1 \le i \le 16$, are generated from the key K. From the 64 bits of K the 56 non-redundant bits are selected in a particular order indicated by the permuted choice PC-1, listed in Table 6.7. So bit 57 goes to position 1, bit 49 to position 2, etc.. The leftmost 28 bits of these 56 bits are fed into C_0, the 28 rightmost into D_0. In each of the subsequent 16 rounds the contents of registers C_{i-1} and D_{i-1}, $1 \le i \le 16$, are shifted cyclically over LS (i) positions to the left and put in registers C_i resp. D_i. In Table 6.8 the values of LS (i), $1 \le i \le 16$, are listed. With the permuted choice PC-2 48 bits are selected to generate K_i.

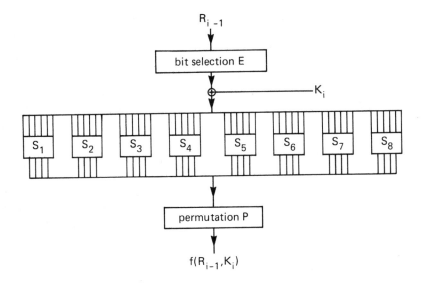

Figure 6.5 Calculation of f (R_{i-1}, K_i)

Many people have criticized the decision to make DES a standard. The two main objections are:

i) The effective keysize (56 bits) is too small for an (hostile) organization with sufficient resources. An exhaustive keysearch is, at least in principle, possible.

ii) The design criteria of the S-boxes are not known. Statistical tests however show that the S-boxes are not completely random. Maybe there is a hidden trapdoor in their structure.

32	1	2	3	4	5
4	5	6	7	8	9
8	9	10	11	12	13
12	13	14	15	16	17
16	17	18	19	20	21
20	21	22	23	24	25
24	25	26	27	28	29
28	29	30	31	32	1

16	7	20	21
29	12	28	17
1	15	23	26
5	18	31	10
2	8	24	14
32	27	3	9
19	13	30	6
22	11	4	25

Bit-selection table E Permutation P

Table 6.6 Tables needed for the calculation of $f(R_{i-1}, K_i)$

Since the publication of the DES algorithm however no effective way of breaking DES has been published. We refer the interested reader to [Des85], [DiH76], [Hel79] and [Mor77].

57	49	41	33	25	17	9
1	58	50	42	34	26	18
10	2	59	51	43	35	27
19	11	3	60	52	44	36
63	55	47	39	31	23	15
7	62	54	46	38	30	22
14	6	61	53	45	37	29
21	13	5	28	20	12	4

14	17	11	24	1	5
3	28	15	6	21	10
23	19	12	4	26	8
16	7	27	20	13	2
41	52	31	37	47	55
30	40	51	45	33	48
44	49	39	56	34	53
46	42	50	36	29	32

PC-1 PC-2

Table 6.7 Permuted choices PC-1 and PC-2.

round i	1	2	3	4	5	6	7	8	9	10	11	12	13	14	15	16
$LS(i)$	1	1	2	2	2	2	2	2	1	2	2	2	2	2	2	1

Table 6.8 $LS(i)$ = # shifts to the left, $1 \le i \le 16$.

		column															
		0	1	2	3	4	5	6	7	8	9	10	11	12	13	14	15
row	0	14	4	13	1	2	15	11	8	3	10	6	12	5	9	0	7
	1	0	15	7	4	14	2	13	1	10	6	12	11	9	5	3	8
S_1	2	4	1	14	8	13	6	2	11	15	12	9	7	3	10	5	0
	3	15	12	8	2	4	9	1	7	5	11	3	14	10	0	6	13
	0	15	1	8	14	6	11	3	4	9	7	2	13	12	0	5	10
	1	3	13	4	7	15	2	8	14	12	0	1	10	6	9	11	5
S_2	2	0	14	7	11	10	4	13	1	5	8	12	6	9	3	2	15
	3	13	8	10	1	3	15	4	2	11	6	7	12	0	5	14	9
	0	10	0	9	14	6	3	15	5	1	13	12	7	11	4	2	8
	1	13	7	0	9	3	4	6	10	2	8	5	14	12	11	15	1
S_3	2	13	6	4	9	8	15	3	0	11	1	2	12	5	10	14	7
	4	1	10	13	0	6	9	8	7	4	15	14	3	11	5	2	12
	0	7	13	14	3	0	6	9	10	1	2	8	5	11	12	4	15
	1	13	8	11	5	6	15	0	3	4	7	2	12	1	10	14	9
S_4	2	10	6	9	0	12	11	7	13	15	1	3	14	5	2	8	4
	3	3	15	0	6	10	1	13	8	9	4	5	11	12	7	2	14
	0	2	12	4	1	7	10	11	6	8	5	3	15	13	0	14	9
	1	14	11	2	12	4	7	13	1	5	0	15	10	3	9	8	6
S_5	2	4	2	1	11	10	13	7	8	15	9	12	5	6	3	0	14
	3	11	8	12	7	1	14	2	13	6	15	0	9	10	4	5	3
	0	12	1	10	15	9	2	6	8	0	13	3	4	14	7	5	11
	1	10	15	4	2	7	12	9	5	6	1	13	14	0	11	3	8
S_6	2	9	14	15	5	2	8	12	3	7	0	4	10	1	13	11	6
	3	4	3	2	12	9	5	15	10	11	14	1	7	6	0	8	13
	0	4	11	2	14	15	0	8	13	3	12	9	7	5	10	6	1
	1	13	0	11	7	4	9	1	10	14	3	5	12	2	15	8	6
S_7	2	1	4	11	13	12	3	7	14	10	15	6	8	0	5	9	2
	3	6	11	13	8	1	4	10	7	9	5	0	15	14	2	3	12
	0	13	2	8	4	6	15	11	1	10	9	3	14	5	0	12	7
	1	1	15	13	8	10	3	7	4	12	5	6	11	0	14	9	2
S_8	2	7	11	4	1	9	12	14	2	0	6	10	13	15	3	5	8
	3	2	1	14	7	4	10	8	13	15	12	9	0	3	5	6	11

Table 6.9 S-boxes (selection functions)

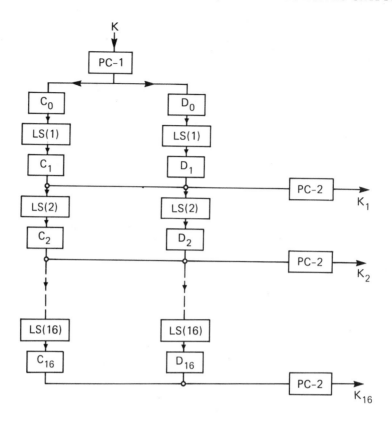

Figure 6.10 Key schedule calculation.

Problems

1. Let S be a plaintext source that generates independent, identical distributed letters X from $\{a, b, c, d\}$. The probability distribution is given by $Pr\ (X=a) = 1/2$, $Pr\ (X=b) = 1/4$, and $Pr\ (X=c) = Pr\ (X=d) = 1/8$.

    Consider the two coding schemes:

a	→	00		a	→	0
b	→	01		b	→	10
c	→	10		c	→	110
d	→	11		d	→	111

    Coding A   Coding B

    The output sequence of the plaintext X is first converted into a $\{0,1\}$-sequence by one of the above coding schemes and subsequently encrypted with the DES algorithm.

    What is the unicity distance for both coding schemes?

7 PUBLIC KEY CRYPTOGRAPHY

In modern day communication systems the conventional cryptosystems turned out to have two disturbing disadvantages.

- The problem of key management and distribution. A communication system with n users, who all use a conventional cryptosystem to communicate with each other, implies the need of $\begin{bmatrix} n \\ 2 \end{bmatrix}$ keys and $\begin{bmatrix} n \\ 2 \end{bmatrix}$ secure channels. Whenever a user wants to change his keys or a new user wants to participate in the system $n - 1$ (resp. n) new keys have to be generated and distributed over as many secure channels.

- The authentication problem. In computer controlled communication systems the electronic equivalence of a signature is needed. Conventional cryptosystems do not provide this feature in a natural way.

These disadvantages prompted researchers to look for a different kind of cryptosystem.

In [Dif76] W. Diffie and M.E. Hellman published their pioneering work on *public key cryptosystems*. See Figure 7.1, where such a system is depicted.

Every user U of the cryptosystem makes his own encryption algorithm E_U and decryption algorithm D_U (or gets them from a trustworthy authority). For reasons that will become clear later, these algorithms must satisfy

$$\text{PK1:} \qquad D_U\left(E_U(m)\right) = m \tag{7.1}$$

for each possible message m and every user U.

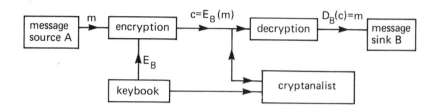

Figure 7.1 A public key cryptosystem.

Every user U makes his encryption algorithm E_U public by putting it in a key book! The decryption algorithm however is kept secret by U. If user A wants to send message m to user B, he first looks up the (public) encryption algorithm E_B of B in the keybook. He encrypts m by applying the algorithm E_B to m. So he sends

$$c = E_B(m). \tag{7.2}$$

User B recovers m from c by applying his (secret) decryption algorithm D_B to the received ciphertext c. Indeed by (7.1)

$$D_B(c) = D_B(E_B(m)) = m. \tag{7.3}$$

For practical resp. cryptographic reasons the encryption and decryption algorithms E_U and D_U must satisfy two more requirements:

> PK2: E_U and D_U are algorithms that do not need much computing time or memory space.

> PK3: It is with all practical means impossible to find an algorithm D_U^* from E_U that satisfies $D_U^*(E_U(m)) = m$ for all possible m.

We summarize this in Table 7.2.

Public	: Encryption algorithm E_U of all users U.
Secret	: Decryption algorithm D_U of all users U.
Properties	: PK1, PK2, PK3.
Encryption	of message m for B
Decryption	by B

Table 7.2 How a public key cryptosystem is used.

It is requirement PK3 that provides the security of the cryptosystem. PK3 makes it possible to publish the encryption algorithms in a keybook without endangering the privacy of the transmitted messages. If a user U wants to change his personal key, he simply makes a new E_U and D_U satisfying PK1, PK2 and PK3 and puts E_U in the keybook. The same has to be done when a new user wants to participate in the communication system. In [Dif76] the authors suggest to use *trapdoor one-way functions* for the encryption. A one-way function is a function f that is easy to evaluate, but whose left inverse f^{-1} is very difficult to compute. A trapdoor one-way function is a one-way function f, whose left inverse f^{-1} is easy to compute given certain additional information. In the next chapters examples of trapdoor one-way functions will be given. One way-functions f are also used to check the authenticity of a person that wants to get access to something. Each user U has his own PIN code x_U, but in the computer only $y_U = f(x_U)$ is stored together with the name of U. When U wants to get access he needs to give his name and x_U. The computer evaluates $f(x_U)$ and checks that $f(x_U) = y_U$. If these two numbers do match, U can get access, otherwise not. The advantage of this system is that the PIN codes x_U do not need to be stored in the computer. So somebody else that can read out the memory of the computer can still not determine the PIN codes. To solve the authentication problem one needs the following requirement instead of PK1.

$$\text{PK4:} \qquad E_U(D_U(m)) = m. \tag{7.4}$$

for all users U and each possible message m.

If user A wants to send message m to B provided with his own signature, he sends

$$c = D_A(m). \tag{7.5}$$

User B recovers m from c by applying the publicly known algorithm E_A to c. Indeed by (7.4)

$$E_A(c) = E_A(D_A(m)) = m. \tag{7.6}$$

The security of this authentication scheme comes from the following requirement:

PK5: It is with all practical means impossible to find an algorithm D_U^* from E_U that satisfies $E_U(D_U^*(m)) = m$ for all possible m.

User A has to make sure that a randomly selected c has a negligible probability of leading to a useful message $E_A(c)$. Of course anybody else can also find the message m from c, so there is no secrecy. But only A can originally have made the pair $(m, D_A(m))$. We summarize this in Table 7.3.

Public	: Encryption algorithm E_U of all users U.
Secret	: Decryption algorithm D_U of all users U.
Properties	: PK2, PK4, PK5.

A sends to B	: $D_A(m) = c$.
B computes	: $E_A(c) = m$.
Signature	: the pair $(m, D_A(m))$.

Table 7.3 How A sends message m with his own signature to B.

If A wants to send message m in encrypted form to B with his own signature attached to it, he needs properties PK1 - PK5. He sends

$$c = E_B (D_A (m)).$$ (7.7)

User B recovers m from c by applying $E_A \circ D_B$ to c. Indeed by (7.1) and (7.4)

$$E_A (D_B (c)) = E_A (D_B (E_B (D_A (m)))) = E_A (D_A (m)) = m.$$

Although everybody can look up E_A in the keybook, it is only B who can recover m from c, because only B knows D_B. B keeps the pair $(m, D_B(c) = D_A(m))$ as A's signature just like above. We summarize this in Table 7.4.

Public	: Encryption algorithm E_U of all users U.
Secret	: Decryption algorithm D_U of all users U.
Properties	: PK1, PK2, PK3, PK4, PK5.
A sends to B	: $E_B (D_A (m)) = c$.
B computes	: $E_A (D_B (c)) = m$.
Signature	: the pair $(m, D_B(c) = D_A(m))$.

Table 7.4: A sends message m with his own signature in ciphertext to B.

In chapters 9, 10 and 11 we shall discuss various proposals for trapdoor one-way functions. In the next chapter we shall meet a one-way function, which is not a trapdoor.

Problems

1. In a communication network every user U has its own public encryption algorithm E_U and private decryption algorithm D_U. A message m from user A to user B will always be send in the format $(E_B (m), A)$. The address A tells B from whom the message originates. Receiver B will retrieve m from $(E_B (m), A)$, but user B will also automatically send $(E_A (m), B)$ back to user A (note that $(E_A (m), B))$ has the right format). In this way A knows that his message has been properly received by B.

    a) Show how a third user C of the network can retrieve message m, that was send by A to B. You may assume that user C can intercept all messages, that are communicated over the network, and that C can also transmit his own texts, as long as they have the right format.

    b) Show that communication over this network is still not safe if the protocol is such that A sends $E_B ((E_B (m), A)$ to B and B automatically sends $E_A ((E_A (m), B)$ back to A.

8 THE DISCRETE LOGARITHM PROBLEM

§ 8.1 The discrete logarithm system

In [Dif76] Diffie and Hellman propose a public key distribution system which is based on the apparent difficulty of computing logarithms over the finite field $GF(p)$, p prime. The reader, who is not familiar with the theory of finite fields is referred to Appendix B. Let α be a primitive element of $GF(p)$. So each nonzero element c in $GF(p)$ can be written as

$$c = \alpha^m, \tag{8.1}$$

where m is unique modulo $p - 1$. If m is given, c can be computed from (8.1) with $2 \cdot \lceil \log_2 p \rceil$ multiplications [Knu69, pp.398-422]. One can realize this by repeated squaring or multiplying by α. The binary representation of m gives the order of these operations. For instance the binary representation 10101011 corresponds in the following way with the computation of α^{171}

$$10 \quad 10 \quad 10 \quad 1 \quad 1$$

$$\alpha^{171} = (((((((\alpha)^2)^2 \alpha)^2)^2 \alpha)^2)^2 \alpha)^2 \alpha.$$

The opposite problem of finding m satisfying (8.1), from c, is not so easy. It is called the *discrete logarithm problem*.

Definition 8.1 Let f and g be functions from **N** into **N**. Then we write

$$f(n) = O(g(n)),$$

if there exists a constant c such that for all $n \in \mathbf{N}$

$$f(n) \leq c \cdot g(n).$$

We shall often use the *O-symbol* , defined above, to give a measure of the number of operations needed in an algorithm. In [Knu73, pp.9, 575-576] one can find an algorithm that solves the logarithm problem. It involves roughly $O(\sqrt{p})$ operations and $O(\sqrt{p})$ bits of memory space. In Theorem 8.6 a more precise analysis of this algorithm will be given. Writing $t = \log_2 p$ one gets the following exponential relation between exponentiation and taking logarithms.

exponentiation   : $O(t)$ operations,

taking logarithms   : $O(2^{t/2})$ operations.

We shall now describe how this discrepancy in computing time can be used to make a public key distribution system. Each user U chooses m_U, $1 \leq m_U \leq p - 2$, at random, computes $\alpha^{m_U} = c_U$ and puts c_U in the public keybook. U keeps m_U secret. Let us now assume that A and B want to communicate with each other using a conventional cryptosystem, but have no secure channel to exchange a key. With the public keybook, they can agree on the secret key

$$k_{A,B} = \alpha^{m_A m_B}. \tag{8.2}$$

A can compute $k_{A,B}$ by raising the publicly known c_B to the power m_A, which only A knows himself. Indeed

$$c_B^{m_A} = (\alpha^{m_B})^{m_A} = \alpha^{m_A m_B} = k_{A,B}.$$

Similarly B finds $k_{A,B}$ by computing $c_A^{m_B}$. If somebody else is able to compute m_A from c_A (or m_B from c_B), he can compute $k_{A,B}$ just like A or B did. By taking p sufficiently large the computation time of solving this logarithm problem will be prohibitively large. Diffie and Hellman suggest to take p about 100 bits long. A different way of finding $k_{A,B}$ from c_A and c_B does not seem to exist. There is no obvious reason to restrict the size of the finite field to a prime number. So from now on the size of the field can be any prime power $q = p^e$. In [Lün87, Chapter XIII] efficient algorithms to find primitive elements in finite fields are described. We summarize the key distribution system in Table 8.1.

Public   : q, a primitive element α in $GF(q)$ and $c_U \ (= \alpha^{m_U})$ of all users U.

Secret   : m_U, only known to user U.

Key used by A and B   : $k_{A,B} = \alpha^{m_A m_B}.$

A computes   : $c_B^{m_A} = k_{A,B}.$

B computes   : $c_A^{m_B} = k_{A,B}.$

Table 8.1: The public key distribution system, based on the discrete logarithm problem.

§ 8.2 How to take discrete logarithms?

The Pohlig-Hellman algorithm

In [Poh78] Pohlig and Hellman demonstrate that discrete logarithms can be taken much faster than in $O(\sqrt{q})$ operations, if $q - 1$ has only small prime divisors. We shall first demonstrate this method for two special cases.

Special Case 8.2: $q - 1 = 2^n$.

The problem is to find m, $0 \le m \le q - 2$, satisfying (8.1), for a given value of c. Let $m_{n-1} \cdots m_1 m_0$ be the binary representation of m, i.e.

$$m = m_0 + m_1 2 + \cdots + m_{n-1} 2^{n-1}, \tag{8.3}$$

$m_i \in \{0, 1\}$, $0 \le i \le n - 1$. Of course it suffices to compute the unknown m_i's. Since α is a primitive element of $GF(q)$, we know that $\alpha^{q-1} = 1$. It also follows that $\alpha^{(q-1)/2} = -1$, because $\alpha^{(q-1)/2} \ne 1$ and $(\alpha^{(q-1)/2})^2 = 1$. Hence

$$c^{\frac{q-1}{2}} = (\alpha^m)^{\frac{q-1}{2}} = \alpha^{(m_0 + m_1 2 + \cdots + m_{n-1} 2^{n-1}) \frac{q-1}{2}} = \alpha^{m_0 \frac{q-1}{2}} = \begin{cases} 1 & \text{if } m_0 = 0, \\ -1 & \text{if } m_0 = 1 \end{cases} \tag{8.4}$$

So the evaluation of $c^{\frac{q-1}{2}}$, which costs at most $2\lceil \log_2 q \rceil$ operations, as we have remarked before, yields m_0. Compute $c \cdot \alpha^{-m_0} = c_1$. Now m_1 can be determined in the same way as above from

$$c_1^{\frac{q-1}{4}} = (\alpha^{m_1 + m_2 2 + \cdots + m_{n-1} 2^{n-2}})^{\frac{q-1}{2}} = \alpha^{m_1 \frac{q-1}{2}} = \begin{cases} 1 & \text{if } m_1 = 0, \\ -1 & \text{if } m_1 = 1. \end{cases} \tag{8.5}$$

Compute $c_1 \cdot \alpha^{-2m_1} = c_2$ and determine m_2 from $(c_2)^{\frac{q-1}{8}}$, etc.. All together this algorithm finds m from c in at most

$$n \cdot (2\lceil \log_2 q \rceil + 2) \text{ operations}, \tag{8.6}$$

where the term 2 comes from the evaluation of c_i, $1 \le i \le n - 1$.

Note that when $q - 1 = s \cdot 2^t$, s odd, the t least significant bits of m can be found in exactly the same way.

Special case 8.3: $q - 1 = p_1^{n_1} p_2^{n_2} \cdots p_k^{n_k}$, $n_i > 0$, $1 \le i \le k$, where all p_i, $1 \le i \le k$, are small, distinct prime numbers.

First $m^{(i)} = (m \bmod p_i^{n_i})$ will be determined for all $1 \le i \le k$. With the Chinese Remainder Theorem (Theorem A.23) m can subsequently be computed from $m^{(i)}$, $1 \le i \le k$, in $O(k \cdot \log_2 q)$ operations and with $O(k \cdot \log_2 q)$ bits of memory space (see Theorem A.24). To determine $m^{(1)}$ (the others can be found in the same way) we write

$$m^{(1)} = m_0 + m_1 p_1 + \cdots + m_{n_1-1} p_1^{n_1-1}, \tag{8.7}$$

$0 \le m_j \le p_1 - 1$, $0 \le j \le n_1 - 1$. Similarly to the previous case we will find m_0 by evaluating $c^{(q-1)/p_1}$. From $(c^{(q-1)/p_1})^{p_1} = 1$, it follows that $c^{(q-1)/p_1}$ is a p_1-th root of unity. Define the primitive p_1-th root of unity β_1 by $\beta_1 = \alpha^{(q-1)/p_1}$ and make a table of $1, \beta_1, \ldots, \beta_1^{p_1-1}$. Then

$$c^{\frac{q-1}{p_1}} = (\alpha^m)^{\frac{q-1}{p_1}} = (\alpha^{m^{(1)}})^{\frac{q-1}{p_1}} = \alpha^{m_0 \frac{q-1}{p_1}} = \beta_1^{m_0}. \tag{8.8}$$

So a simple table lookup of $c^{\frac{q-1}{p_1}}$ will yield m_0. To determine m_1 first compute $c_1 = c \cdot \alpha^{-m_0}$ and then evaluate $c_1^{(q-1)/p_1^2}$. Etc.. For this algorithm we have to calculate the values of $1, \beta_i, \ldots, \beta_i^{p_i-1}$, $1 \le i \le k$, and put them in a table. Each time that we want to take a logarithm the algorithm will have to take $\sum_{i=1}^{k} n_i$ exponentiations. So the algorithm involves

$$\sum_{i=1}^{k} 2 \cdot \lceil \log_2 q \rceil \cdot n_i \approx 2 \cdot (\sum_{i=1}^{k} n_i) \cdot \log_2 q \le 2 \cdot (\log_2 q)^2 \tag{8.9}$$

operations, if we forget about the lower order terms.

Example 8.4: $q = 8101$, $q - 1 = 2^2 3^4 5^2$, $\alpha = 6$, $\alpha^{-1} = 6751$.

We first perform some precalculations. With the primitive 2-nd, 3-rd and 5-th roots of unity

$$\begin{aligned}
\beta_1 &= \alpha^{(8101-1)/2} &= \alpha^{4050} &= 8100, \\
\beta_2 &= \alpha^{(8101-1)/3} &= \alpha^{2700} &= 5883, \\
\beta_3 &= \alpha^{(8101-1)/5} &= \alpha^{1620} &= 3547,
\end{aligned}$$

we make the following three tables:

$p_1 = 2$:

i	0	1
β_1^i	1	8100

$p_2 = 3$:

i	0	1	2
β_2^i	1	5883	2217

$p_3 = 5$:

i	0	1	2	3	4
β_3^i	1	3547	356	7077	5221

From the Chinese Remainder Theorem (Theorem A.23) one gets:

$$\left. \begin{array}{l} a \equiv 1 \bmod 4 \\ a \equiv 0 \bmod 81 \\ a \equiv 0 \bmod 25 \end{array} \right\} \Rightarrow a \equiv 2025 \bmod 8100 \; ,$$

$$\left. \begin{array}{l} b \equiv 0 \bmod 4 \\ b \equiv 1 \bmod 81 \\ b \equiv 0 \bmod 25 \end{array} \right\} \Rightarrow b \equiv 6400 \bmod 8100$$

and

$$\left. \begin{array}{l} c \equiv 0 \bmod 4 \\ c \equiv 0 \bmod 81 \\ c \equiv 1 \bmod 25 \end{array} \right\} \Rightarrow c \equiv 7776 \bmod 8100 \; .$$

This concludes the preliminary work.

Let us now solve (8.1) for $c = 7531$. We first determine $m^{(i)} = (m \underline{\bmod} \; p_i^{n_i})$, $1 \le i \le 3$, with the method explained above.

$p_1 = 2$, $n_1 = 2$:

c	$= 7531$,	$c^{(8101-1)/2}$	$= 8100$,	$m_0 = 1$,
$c_1 = \quad c \cdot \alpha^{-1}$	$= 8006$,	$c_1^{(8101-1)/2^2}$	$= \quad 1$,	$m_1 = 0$.

Hence $m^{(1)} = 1 + 0.2^1 = 1$.

$p_2 = 3$, $n_2 = 4$:

c	$= 7531$,	$c^{(8101-1)/3}$	$= 2217$,	$m_0 = 2$,
$c_1 = \quad c \cdot \alpha^{-2}$	$= 6735$,	$c_1^{(8101-1)/3^2}$	$= \quad 1$,	$m_1 = 0$,
$c_2 = \quad c_1$	$= 6735$,	$c_2^{(8101-1)/3^3}$	$= 2217$,	$m_2 = 2$,
$c_3 = \quad c_2 \cdot \alpha^{-2 \cdot 3^2}$	$= 6992$,	$c_3^{(8101-1)/3^4}$	$= 5883$,	$m_3 = 1$.

Hence $m^{(2)} = 2 + 0.3^1 + 2.3^2 + 1.3^3 = 47$.

$p_3 = 5$, $n_3 = 2$:

c	$= 7531$,	$c^{(8101-1)/5}$	$= 5221$,	$m_0 = 4$,
$c_1 = \quad c \cdot \alpha^{-4}$	$= 7613$,	$c_1^{(8101-1)/5^2}$	$= \quad 356$,	$m_1 = 2$.

Hence $m^{(3)} = 4 + 2.5^1 = 14$.

The solution m is now given by

$$\begin{aligned} m &\equiv \quad a \cdot m^{(1)} + b \cdot m^{(2)} + c \cdot m^{(3)} \equiv \\ & \quad 2025 \times 1 + 6400 \times 47 + 7776 \times 14 \equiv 6689 \bmod 8100. \end{aligned}$$

If $q - 1$ has large prime factors, the dominant term in the workload will either be the making of the tables (and the table lookup) or the repeated, direct computation of the exponents m_j, $1 \leq j \leq n_i$, $1 \leq i \leq k$. The following lemma treats the complexity of finding the exponents $m^{(i)}$, $1 \leq i \leq k$.

Lemma 8.5 Let α be a primitive element of $GF(q)$. Let p be a divisor of $q - 1$ (not necessarily prime) and define $\beta = \alpha^{(q-1)/p}$. So β is a primitive p-th root of unity. Let c be any p-th root of unity. Then for every t, $0 \leq t \leq 1$, one can find m, $0 \leq m \leq p - 1$ satisfying

$$c = \beta^m. \tag{8.10}$$

with an algorithm that uses

$$O(p^{1-t}(1 + \log_2 p^t)) \text{ operations,} \tag{8.11}$$

$$O(p^t \cdot \log_2 q) \text{ bits of memory space} \tag{8.12}$$

and one initial precalculation involving

$$O(p^t \cdot (1 + \log_2 p^t)) \text{ operations.} \tag{8.13}$$

Proof: Let $s = \lceil p^t \rceil$. Define d and r by

$$m = d \cdot s + r, \tag{8.14}$$

$$0 \leq r < s \approx p^t. \tag{8.15}$$

Observe that

$$0 \leq d \leq m/s \leq p/s \approx p^{1-t}. \tag{8.16}$$

Of course solving (8.10) is equivalent to finding d and r satisfying (8.15) and

$$\beta^r = c \cdot \beta^{-ds}. \tag{8.17}$$

We start with the initial precalculation. First we make a table of $1, \beta, \ldots, \beta^{s-1}$. This requires $s \approx p^t$ operations. Then we sort this table in $O(p^t \cdot \log_2 p^t)$ operations [Knu73, pp.184]. Together this explains (8.13). Each of the $s \approx p^t$ field elements needs $\log_2 q$ bits of memory space. This explains (8.12). To solve (8.17) we compute $c \cdot \beta^{-is}$, $i = 0, 1, \ldots, \lfloor p/s \rfloor$, and look for this element in the table. At $i = d$ we will find a match and we are done. For each value i we have to perform 1 multiplication and a table lookup, which costs $O(\log_2 p^t)$ operations. It follows from (8.16) that we have to try at most $O(p^{1-t})$ possible values of i. This explains (8.11). □

Theorem 8.6 Let α be a primitive element in $GF(q)$, where $q - 1 = p_1^{n_1} p_2^{n_2} \cdots p_k^{n_k}$ with $p_1 < p_2 < \cdots < p_k$ prime and $n_i > 0$, $1 \leq i \leq k$. For any sequence $\{t_i\}_{i=1}^k$ with $0 \leq t_i \leq 1$, $1 \leq i \leq k$, one can solve m, $0 \leq m \leq q - 2$, from

$$c = \alpha^m \tag{8.18}$$

with an algorithm that uses

$$O \left(\sum_{i=1}^{k} \{ n_i [2 \log_2 q + p_i^{1-t_i} \cdot (1 + \log_2 p_i^{t_i})] \} \right) \text{ operations,} \tag{8.19}$$

$$O \left(\sum_{i=1}^{k} \{ 1 + p_i^{t_i} \} \cdot \log_2 q \right) \text{ bits of memory space} \tag{8.20}$$

and one initial precalculation involving

$$O \left(\sum_{i=1}^{k} \{ \log_2 q + p_i^{t_i} \cdot (1 + \log_2 p_i^{t_i}) \} \right) \text{ operations.} \tag{8.21}$$

Proof: As before we first compute $m^{(i)} = (m \mod p_i^{n_i})$, $1 \le i \le k$. From these one can find m with the Chinese Remainder Theorem (see Theorem A.23). This involves a precomputation of $O (k \cdot \log_2 q)$ operations and $O (k \cdot \log_2 q)$ bits of memory space. The final computation of m from the $m^{(i)}$'s only takes $k < \log_2 q$ multiplications (see Theorem A.24). This explains the first term in (8.20) and the first term in (8.21). To determine $m^{(i)}$, $1 \le i \le k$, we follow the strategy explained in Special case 8.3. So for each $1 \le i \le k$ we write

$$m^{(i)} = m_0 + m_1 p_1 + \cdots + m_{n_i - 1} p_i^{n_i - 1}.$$

The computation of each m_j, $0 \le j \le n_i - 1$, involves one exponentiation and one invocation of Lemma 8.5. This accounts for the two terms in (8.19) and the second terms in (8.20) and (8.21).   []

For $t_1 = t_2 = \cdots = t_k = \frac{1}{2}$ this algorithm reduces to the $O (\sqrt{q})$ algorithm that was mentioned at the beginning of § 8.1 [Knu73, pp.9]. Note that the product of computing time and bits of memory space in the above algorithm is more or less constant.

The Adleman algorithm

In [Adl79] Adleman describes an algorithm that computes a logarithm over the field $GF (q)$, q prime, using

$$O (\exp(C \sqrt{\ln q \, \ln \ln q})) \tag{8.22}$$

operations, where C is a constant. In [Hel83] it is shown that this method can be extended to all fields $GF (q)$. We give a rough outline of this algorithm for the case $q = 2^n$.

Let $GF (2^n) = GF (2)[x]/(p (x))$, where $p (x)$ is a primitive, binary polynomial of degree n (see Appendix B). The elements of $GF (2^n)$ can be represented as binary polynomials in x of degree $< n$. The field element x is the primitive element α of (8.1). The field element c is in this terminology a binary polynomial $c (x)$ of degree $< n$. Choose $b \approx C' \sqrt{n \ln n}$, where C' is a small constant. A polynomial $a (x)$ is called *smooth* with respect to b if $a (x)$ is the product of binary, irreducible polynomials of degree $\leq b$. Let $q_i(x)$, $1 \leq i \leq L$, be a list of all binary, irreducible polynomials of degree $\leq b$. Then $L \approx 2^b / b$ (see Corollary B.23). Choose m, $1 \leq m \leq q - 2$, random and set $a(x) = x^m \underline{\mod} p(x)$, which analogous to (2.1) means that $a(x) \equiv x^m \mod p(x)$, degree $(a(x)) < n$. Test whether $a(x)$ is smooth with respect to b (divide $a(x)$ by the various polynomials in the list). If $a(x)$ is not smooth with respect to b, discard it and try a new choice of m. If $a(x)$ is smooth, there is a unique integer vector $\underline{e} = (e_1, e_2, \ldots, e_L)$, $e_i \geq 0$, $1 \leq i \leq L$, such that

$$x^m \underline{\mod} p(x) = a(x) = \prod_{i=1}^{L} (q_i(x))^{e_i}. \tag{8.23}$$

Define the integer vector $\underline{r} = (r_1, r_2, \ldots, r_L)$, $0 \leq r_i < q - 1$, $1 \leq i \leq L$, by

$$q_i(x) \equiv x^{r_i} \mod p(x), \quad 1 \leq i \leq L. \tag{8.24}$$

For the computation of the r_i's one would have to solve the discrete logarithm problem. So at this moment we have to regard them as unknown quantities. On the other hand (8.23) gives the following congruence relation

$$m \equiv \sum_{i=1}^{L} e_i \cdot r_i \mod q - 1. \tag{8.25}$$

If one has L independent linear equations of the type (8.25), one can solve $\underline{r}$. So we repeat the above procedure as often as is needed to get L linearly independent congruence relations of the type (8.25). The expected number of congruence relations that are needed to determine $\underline{r}$ will be slightly more than $L \approx 2^b / b$. This concludes the initial precalculation. In [Odl84] one can find a proof that the fraction of the polynomials of degree $< n$ that are smooth with respect to b is about

$$(n/b)^{-n/b}. \tag{8.26}$$

So the expected number of trials before the above choice of m gives rise to a polynomial $a(x)$ that is smooth with respect to b, is the reciprocal of (8.26). Each test involves $L \approx 2^b / b$ divisions. So the number of operations in this initial precalculation is about

$$\left[\frac{2^b}{b}\right]^2 \cdot \left[\frac{n}{b}\right]^{n/b}. \tag{8.27}$$

Now consider an arbitrary field element $c(x)$. To solve (8.1) we choose a random integer $0 \leq j \leq q - 2$ and test whether $c(x) \cdot x^j \underline{\mod} p(x)$ is smooth with respect to b. If so, one can find the solution m from the factorization of $c(x) \cdot x^j \underline{\mod} p(x)$ (in polynomials $q_i(x)$, $1 \leq i \leq L$), and the vector $\underline{r}$ that contains the logarithms of these polynomials $q_j(x)$. If $c(x) \cdot x^j \mod p(x)$ is not smooth, we try again with a new choice of j and repeat the above procedure until a polynomial is found that is smooth with respect to b. The workfactor of this part is given by

$$2^b / b \cdot (n/b)^{n/b}, \tag{8.28}$$

which is the product of the expected number of choices of j that one has to try times the workfactor of the test for smoothness of $c(x) \cdot x^j \underline{\mod} p(x)$.

Substitution of $b = C' \sqrt{n \ln n}$ in (8.28) (as in (8.27)) yields (8.22).

In [Bla84] the authors propose an algorithm that reduces the probabilistic aspect of the Adleman algorithm. Allthough this algorithm is faster than Adleman's algorithm, its asymptotic running time is the same.

Coppersmith [Cop84] describes an algorithm for the case $q = 2^n$ with asymptotic running time

$$O\ (\exp(C\ n^{1/3} \ln^{2/3} n\ )). \tag{8.29}$$

This implies that for q a power of 2, one will have to take a larger value of q to obtain the same security as an odd value of q would give.

For further reading we refer the reader to the excellent survey on the discrete logarithm problem by Odlyzko [Odl85].

Problems

1. Users A and B want to use the discrete logarithm system to fix a common key over a public channel. They use $F_2[x]/(x^{10}+x^3+1)$ as representation of GF (2^{10}). User B makes $c_B = 0100010100$ public, which stands for the field element $x + x^5 + x^7$. If $m_A = 2$, what will be the common key that A and B use for their communication?

2. Demonstrate the Special Case 8.2 version of the Pohlig-Helmann algorithm, that computes logarithms in finite fields of size $q = 2^n + 1$, by evaluating $\log_3 (15)$ in GF(17).

3. Compute $\log_3 (135)$ in GF(353) with the Pohlig-Hellman algorithm.

9 RSA

§ 9.1 The RSA system

In 1978 R.L. Rivest, A. Shamir and L. Adleman [Riv78] proposed a public key cryptosystem that has become known as the *RSA system,* although one sometimes also sees the name "the MIT system". It depends on Euler's Theorem (see Definition A.13 and Theorem A.16).

Theorem 9.1 (Euler)
Let a and n be integers. Then

$$gcd(a,n) = 1 \qquad \Rightarrow \qquad a^{\phi(n)} \equiv 1 \bmod n, \tag{9.1}$$

where *Euler's Totient Function* ϕ is defined by, resp. satisfies

$$\phi(n) = \left| \{1 \le i \le n \mid gcd(i,n) = 1\} \right| = n \cdot \prod_{p \mid n} (1 - 1/p). \tag{9.2}$$

Each user U chooses two different primes, say p_U and q_U, of about 100 digits long. Let $n_U = p_U \cdot q_U$. It follows from (9.2) that

$$\phi(n_U) = (p_U - 1)(q_U - 1). \tag{9.3}$$

Secondly U chooses an integer $1 < e_U < \phi(n_U)$ with $gcd(e_U, \phi(n_U)) = 1$. With Euclid's Algorithm (see § A.2) U computes in less than $2 \cdot \log_2 \phi(n_U)$ operations the integer d_U, satisfying

$$e_U \cdot d_U \equiv 1 \bmod \phi(n_U), \quad 1 < d_U < \phi(n_U). \tag{9.4}$$

Each user U publishes e_U and n_U, but keeps d_U secret. The primes p_U and q_U do no longer play a role (but can not be made public). If user A wants to send a message to user B, he represents his message in any standard way by numbers $0 < m < n_B$. User A looks up e_B in the public keybook and sends the ciphertext

$$c = m^{e_B} \underline{\bmod}\ n_B . \tag{9.5}$$

User B can recover m from c by computing $c^{d_B} \bmod n_B$. Indeed for some integer l one has by (9.4) and (9.1)

$$c^{d_B} \equiv m^{e_B d_B} \equiv m^{1+l\phi(n_B)} \equiv m \bmod n_B, \tag{9.6}$$

when $gcd(m, n_B) = 1$. We summarize this system in Table 9.1. We invite the reader to verify that this system also works when $gcd(m, n_B) \neq 1$.

Public	: e_U and n_U of all users U.
Secret	: d_U of user U.
Property	: $e_U \cdot d_U \equiv 1 \bmod \phi(n_U)$.
Messages from A to B	: $0 < m < n_B$.
Encryption by A	: $c = m^{e_B} \underline{\bmod}\ n_B$.
Decryption by B	: $c^{d_B} \underline{\bmod}\ n_B = m$.

Table 9.1: The RSA system without signature.

A cryptanalist can compute m from c in exactly the same way as B, once he knows the secret d_B. He is able to compute d_B from (9.4) with Euclid's Algorithm and the publicly known e_B, if he knows $\phi(n_B)$. To find $\phi(n_B)$ from (9.3) and the publicly known n_B, a cryptanalist has to find the factorization of n_B. Schroeppel (not published) has a modification of the factorization algorithm by Morrison and Brillhart [MoB75]. It needs

$$O\left(\exp(\sqrt{\ln n \ln \ln n})\right) \text{ operations.} \tag{9.7}$$

The similarity between (9.7) and (8.22) is no coincidence. The interested reader is invited to analyse and compare the two algorithms. Table 9.2 gives an impression of the growth of the expression in (9.7). If n is about 200 digits long, the above cryptanalysis is clearly not feasible. Another fast factorization algorithm can be found in [Len86]. There do exist special factorization algorithms that run faster, if the factorization of n is of a special form. We shall discuss these methods later in this paragraph.

# digits in n	# operations needed to factor n
50	$1.42 * 10^{10}$
75	$8.99 * 10^{12}$
100	$2.34 * 10^{15}$
125	$3.41 * 10^{17}$
150	$3.26 * 10^{19}$
175	$2.25 * 10^{21}$
200	$1.20 * 10^{23}$

Table 9.2: Growth of $\exp(\sqrt{\ln n \, \ln \ln n})$.

With the RSA system one can write signatures. Indeed by (9.4)

$$(m^{d_U})^{e_U} \equiv m^{e_U d_U} \equiv m \bmod n_U \qquad (9.8)$$

for all users U and all $0 < m < n_U$. Table 9.3 summarizes the signature scheme. See also Table 7.3 and the corresponding discussion in Chapter 7.

Public	: e_U and n_U of all users U.
Secret	: d_U by user U.
Property	: $e_U \cdot d_U \equiv 1 \bmod \phi(n_U)$.
Messages from A to B	: $0 < m < n_A$.
m with signature by A	: $c = m^{d_A} \bmod n_A$.
B computes	: $c^{e_A} \bmod n_A = m$.
Signature	: the pair $(m, c = m^{d_A} \bmod n_A)$.

Table 9.3: The RSA system as signature scheme.

Suppose that user A wants to send message m in ciphertext and with his own signature to user B. To make the mapping $m \to (m^{d_A} \bmod n_A)^{e_B} \bmod n_B$ (see (7.7)) one-to-one, one needs $n_A \leq n_B$. Indeed suppose that $n_A > n_B$. Then $m^{d_A} \bmod n_A$ may very well be more than n_B. As a consequence there are different messages that are mapped into the same c. If B also wants to send a message to A in the above way, one needs similarly that $n_A \geq n_B$. So $n_A = n_B$. This greatly reduces the security of the system. We shall now discuss two methods to solve this problem. The *first method* is explained in Table 9.4.

	$n_A < n_B$	$n_A > n_B$
A sends	$c = (m^{d_A} \underline{\bmod} \ n_A)^{e_B} \underline{\bmod} \ n_B$	$c = (m^{e_B} \underline{\bmod} \ n_B)^{d_A} \underline{\bmod} \ n_A$
B computes	$(c^{d_B} \underline{\bmod} \ n_B)^{e_A} \underline{\bmod} \ n_A = m$	$(c^{e_A} \underline{\bmod} \ n_A)^{d_B} \underline{\bmod} \ n_B = m$
B gives to	m and $x = c^{d_B} \underline{\bmod} \ n_B$	m and c
arbiter	$= m^{d_A} \underline{\bmod} \ n_A$	
Arbiter com-	$m' = x^{e_A} \underline{\bmod} \ n_A$	$x = m^{e_B} \underline{\bmod} \ n_B$ and
putes		$x' = c^{e_A} \underline{\bmod} \ n_A$
Arbiter checks	$m = m'$	$x = x'$

Table 9.4 RSA for privacy and authentication (method 1).

If there is an argument between users A and B, they will go to an arbiter. This arbiter proceeds as indicated in Table 9.4. If $m = m'$ resp. $x = x'$ the arbiter will decide that the message indeed came from A. In the other case the arbiter will decide the opposite. Note that the arbiter does not have to know the secret d_A or d_B to make his decision. So A and B can continue to use their original set of parameters.

The *second method* lets every user U choose two sets of parameters, say n_{U_i}, e_{U_i} and d_{U_i}, $i = 1, 2$, with

$$n_{U_1} < 10^{200} < n_{U_2} \tag{9.9}$$

If A wants to send a message to B, he will use n_{A_1}, e_{A_1} and d_{A_1} as his set of parameters and n_{B_2}, e_{B_2} and d_{B_2} as B's set of parameters. They now proceed as described in the first column of Table 9.4.

We shall now briefly discuss a few factorization algorithms that for certain values of n are faster than (9.7). In [Pol75] Pollard describes a way to factor n in $O(\sqrt{p})$ steps, where p is the smallest prime divisor of n. This explains why we have to take p and q both large. On the other hand $|p - q|$ should also be large, because

$$4n = (p + q)^2 - (p - q)^2. \tag{9.10}$$

If one could guess $p - q$, one could find $p + q$ from (9.10) and subsequently solve p and q. The system would then have been broken. So $|p - q|$ has also to be large. One can easily make $|p - q|$ of about the same size as p and q, by taking p a few digits longer in length than q. In [Wlm79] some algorithms are discussed that factor n faster, if one of the integers $p - 1, p + 1, q - 1$ or $q + 1$ has only small prime factors.

In the literature one can also find a few attacks on the RSA system, that have a probability of success which is not significantly more than the probability that a randomly chosen integer $1 < i < n$ has $gcd(i, n) \neq 1$. The latter attack would of course also give p (or q), but its probability of success is

$$\frac{n - \phi(n)}{n} = \frac{p + q - 1}{p \cdot q} \approx \frac{1}{p} + \frac{1}{q} \approx \frac{1}{10^{100}}. \tag{9.11}$$

Because these "attacks" have such a small probability of success, we choose not to discuss them here. To make the RSA system practical, there has to be an efficient way to generate prime numbers of 100 digits long. The following pseudo-algorithm describes how this can be done.

Step 1   : Write down a random, odd integer of 100 digits long.
Step 2   : Test this integer for primality. If it is not prime, go back to step 1, otherwise you are done.

Algorithm 9.5: How to generate a prime number.

In the next two paragraphs we shall discuss two ways to test an integer for primality. The first of the two does not guarantee the primality, but the probability that the presumed prime number is actually not prime can be made arbitrary small. The second test (of which only an outline will be given in § 9.3) can guarantee the primality, but is slower. For other test we refer the reader to [Knu81, § 4.5.4]. How often do we expect to have to go through steps 1 and 2 in the above "algorithm" before obtaining a prime? To answer this question we have to know the fraction of the prime numbers in the set of odd, 100-digit numbers. To this end we quote the Prime Number Theorem [Har45, pp.9]. Let $\pi(x)$ be defined by

$$\pi(x) = \left| \{ 1 \leq p \leq x \mid p = \text{prime} \} \right| \tag{9.12}$$

Theorem 9.2 (Prime Number Theorem)

$$\pi(x) \approx x/\ln x \qquad (x \to \infty). \tag{9.13}$$

With (9.13) one obtains the following approximation of the fraction of odd, 100-digit numbers that are prime.

$$\frac{\pi(10^{100}) - \pi(10^{99})}{\# \text{ odd, 100-digit numbers}} \approx \frac{(10^{100}/\ln 10^{100}) - (10^{99}/\ln 10^{99})}{(10^{100} - 10^{99})/2} =$$

$$\frac{(10/100 \ln 10) - (1/99 \ln 10)}{(10 - 1)/2} = \frac{890}{(9/2) \cdot 100 \cdot 99 \cdot \ln 10} = \frac{890}{44550 \ln 10} \approx \frac{1}{115}.$$

So the expected number of primality tests that one has to perform in the above algorithm is 115. Several people have studied the *bit security* of the RSA system. The bit security expresses the security of one bit of plaintext, given the ciphertext (for instance the least significant bit). In [ChG85] the authors show that it is roughly as hard to determine one bit of plaintext from the ciphertext as the whole plaintext. See also [GMT82].

§ 9.2 The Solovay and Strassen primality test

Let p be a prime. Then an integer a with $p \nmid a$ (read: p does not divide a), is called a *quadratic residue* (QR) modulo p, if the equation

$$x^2 \equiv a \mod p, \tag{9.14}$$

has an integer solution. If $p \nmid a$ and (9.14) does not have an integer solution, a will be called a *quadratic non-residue* modulo p (NQR). The well known *Legendre symbol* (a/p) is defined by

$$(a/p) \equiv \begin{cases} 1 & \text{if } p \nmid a \text{ and } a \text{ is a quadratic residue mod } p, \\ 0 & \text{if } p \mid a, \\ -1 & \text{if } p \nmid a \text{ and } a \text{ is a quadratic non-residue mod } p. \end{cases} \tag{9.15}$$

The *Jacobi symbol* (a/m) generalizes the Legendre symbol to all odd integers m. Let $m = \Pi_i \, p_i$, where the p_i's are (not necessarily distinct) odd primes. Then (a/m) is defined by

$$(a/m) = (a/\Pi_i p_i) = \Pi_i (a/p_i), \tag{9.16}$$

We refer the reader, who is not familiar with this theory to § A.4, where the following rules can be found.

$$(a/m) = ((a-m)/m). \tag{9.17}$$

$$(ab/m) = (a/m)\,(b/m). \tag{9.18}$$

$$(a/mn) = (a/m)\,(a/n). \tag{9.19}$$

$$(a/m)\,(m/a) = (-1)^{(a-1)(m-1)/4}, \text{ for odd coprime integers } a \text{ and } m. \tag{9.20}$$

$$(2/m) = 1 \quad \text{iff} \quad m \equiv \pm 1 \mod 8. \tag{9.21}$$

$$(-1/m) = 1 \quad \text{iff} \quad m \equiv 1 \mod 4. \tag{9.22}$$

With these rules one can evaluate (a/m) very efficiently, as is demonstrated in the following example.

Example 9.3: To find out whether 12703 is a quadratic residue modulo the prime 16361, we compute $(12703/16361)$ with (9.17)-(9.22).

$$(12703/16361) \overset{(9.20)}{=\!=\!=\!=} (16361/12703) \overset{(9.17)}{=\!=\!=\!=} (3658/12703) \overset{(9.18)}{=\!=\!=\!=}$$

$$(2/12703)\,(1829/12703) \overset{(9.20\ \&\ 21)}{=\!=\!=\!=\!=\!=} (12703/1829) \overset{(9.17)}{=\!=\!=\!=} (1729/1829) =\!=$$

$$\overset{(9.20)}{=\!=\!=\!=} (1829/1729) \overset{(9.17)}{=\!=\!=\!=} (100/1729) \overset{(9.18)}{=\!=\!=\!=} (10/1729)^2 = 1.$$

We conclude that 12703 is a quadratic residue mod 16361. Finding an integer solution of $x^2 \equiv 12703$

mod 16361 is an entirely different matter. See [Per86] for a discussion of this problem.

It is easy (see Theorem A.30) to prove that for p prime, $p > 2$ and $p \nmid a$

$$(a/p) \equiv a^{(p-1)/2} \bmod p. \tag{9.23}$$

The reader should recall from the beginning of §8.1, that the right hand side of (9.23) can be evaluated very efficiently too.

The Solovay and Strassen Algorithm [Sol77] relies on the following theorem.

Theorem 9.4: Let m be an odd integer. Define

$$G = \{1 \le a \le m \mid gcd(a, m) = 1 \ \& \ a^{(m-1)/2} \equiv (a/m) \bmod m \}. \tag{9.24}$$

Then

$$|G| = m - 1 \qquad \text{if } m \text{ is prime,} \tag{9.25}$$

$$|G| \le (m-1)/2 \qquad \text{if } m \text{ is not prime.} \tag{9.26}$$

Proof: If m is prime, (9.23) implies that $|G| = m - 1$, i.e. (9.25). So let us assume that m is not a prime. Clearly G is a subgroup of the multiplicative group

$$Z_m^* = \{1 \le i \le m - 1 \mid gcd(i, m) = 1\}. \tag{9.27}$$

It follows that $|G|$ divides $|Z_m^*|$. If $G \ne Z_m^*$ we can conclude that $|G| \le |Z_m^*|/2 = \phi(m)/2 \le$ $\le (m - 1)/2$. This would prove the Theorem. So it suffices to prove the existence of an element a in $\mathbb{Z}_m^*$ which does not satisfy

$$a^{(m-1)/2} \equiv (a/m) \bmod m. \tag{9.28}$$

We distinguish two cases. In [Sol77] the authors forget to consider the case that m is a square. In the proof below, which is due to J.W. Nienhuys, case 1 covers this possibility.

Case 1: $m = p^r \cdot s$ with p an odd prime, $r \ge 2$ and $p \nmid s$. Let a be a solution of

$$a \equiv 1 + p \bmod p^r, \tag{9.29}$$

$$a \equiv 1 \bmod s. \tag{9.30}$$

By the Chinese Remainder Theorem (Theorem A.23) a exists and is unique modulo m. Clearly $ggd(a, p^r) = ggd(a, s) = 1$, so $a \in \mathbb{Z}_m^*$. By (9.29) and the binomial theorem $a^m \equiv (1+p)^m \equiv$ $\equiv 1 \bmod p^r$. By (9.30) $a^m \equiv 1 \bmod s$. It follows from the Chinese Remainder Theorem that $a^m \equiv 1 \bmod m$. Since $a \not\equiv 1 \bmod m$ by (9.29), it also follows that $a^{m-1} \not\equiv 1 \bmod m$. This in turn implies that

$$a^{(m-1)/2} \not\equiv \pm 1 \bmod m .$$ (9.31)

So s cannot satisfy (9.28).

Case 2: $m = p_1 p_2 \cdots p_s$, $s \geq 2$, where the p_i's are distinct prime numbers.
Let u be a quadratic non-residue modulo p_1. By the Chinese Remainder Theorem there is a unique integer a modulo m satisfying

$$a \equiv u \bmod p_1 ,$$ (9.32)

$$a \equiv 1 \bmod p_i , \qquad 2 \leq i \leq s .$$ (9.33)

Clearly $ggd(a,p_i) = 1$, $1 \leq i \leq s$, so $a \in \mathbb{Z}_m^*$. By (9.16) and our assumption on u, we have that $(a/m) = -1$. In particular $(a/m) \equiv -1 \bmod p_i$ for any $1 \leq i \leq s$. On the other hand (9.33) implies that $a^{(m-1)/2} \equiv 1 \bmod p_i$ for any $2 \leq i \leq s$. Hence

$$(a/m) \not\equiv a^{(m-1)/2} \bmod p_i$$ (9.34)

for any i, $2 \leq i \leq s$, and a fortiori (9.28) does not hold.   []

We can now explain the Solovay and Strassen Algorithm.

> prime = true
> for $i = 1$ to k do
> begin   Select a random integer $1 < a < m$;
>   if $gcd(a,m) \neq 1$ or $a^{(m-1)/2} \not\equiv (a/m) \bmod m$ then prime = false
> end.

Algorithm 9.6: The probabilistic primality test by Solovay and Strassen
(k can be any positive integer).

In the algorithm k can be any positive integer. The probability that k independently and randomly selected elements a all will pass the two tests, given in Algorithm 9.6, while m is not a prime, is less than or equal to 2^{-k} by Theorem 9.4. By taking k sufficiently large, the probability that a non-prime number survives the above algorithm can be made arbitrary small.

§ 9.3 The Cohen and Lenstra primality test

At this moment the fastest primality test that proves the (non-)primality of an integer n is due to H. Cohen and H.W. Lenstra jr. [Coh82]. This test is an improvement of [Adl83]. We shall not give a complete description of this test. It involves too much advanced and deep number theory. We shall only give an impression of the algorithm. We closely follow the excellent introductory article by Lenstra [Len83].

Fermat's Theorem (see Theorem A.17), which follows from Theorem 9.1, can be formulated as follows.

Theorem 9.5 (Fermat) Let n be a prime. Then for all integers a

$$a^n \equiv a \bmod n. \tag{9.35}$$

Let n be an integer. A single integer a that does not satisfy (9.35), proves that n is not a prime number. Unfortunately the opposite is not true. For instance $n = 561$ satisfies (9.35), while $n = 3 \times 11 \times 17$. Indeed since $lcm(\phi(3), \phi(11), \phi(17)) = lcm(2, 10, 16) = 80$ (see § A.3), it follows from Euler's Theorem (Theorems 9.1 or A.16)) and the Chinese Remainder Theorem (Theorem A.23) that $a^{80} \equiv 1 \bmod 561$ and a fortiori $a^n \equiv a \bmod n$ for all integers a with $gcd(a, n) = 1$. For the remaining values of a, (9.35) can be proved in a similar way. The converse of a slightly stronger statement than Theorem 9.5 does hold however.

Theorem 9.6 An integer n is prime iff for all integers a

$$gcd(a, n) = 1 \quad \Rightarrow \quad a^{(n-1)/2} \equiv (a/n) \bmod n. \tag{9.36}$$

Proof: That (9.36) holds for prime numbers was already remarked in (9.23). The converse was first proved by Lehmer [Leh76], but it also follows from the proof of Theorem 9.4. ▯

Theorem 9.6 is of course not a very efficient primality test for 100-digit long numbers. Lenstra offers the following "attractive" alternative.

Theorem 9.7 (Lenstra) An integer n is prime iff every divisor d of n is a power of n.

The proof is completely trivial, since $d = 1 = n^0$ and $d = n = n^1$ are the only divisors of a prime number n. Clearly it is not Theorem 9.7 that we want to use as a primality test, but a variant of it does turn out to be very powerful. We shall show that under certain conditions every divisor of n looks a little bit like a power of n.

Theorem 9.8 Let n be any integer with $2 \nmid n$ and $3 \nmid n$. Assume that

    i) $i^{(n-1)/2} \equiv (i/n) \bmod n$ for $i = -1, 2$ and 3,   (9.37)

    ii) $\exists_a [a^{(n-1)/2} \equiv -1 \bmod n]$.   (9.38)

Then

$$\forall_{d \mid n} \exists_j [d \equiv n^j \bmod 24]. \tag{9.39a}$$

In fact (9.39a) can be strengthened to

$$\forall_{d \mid n} \exists_{j \in \{0,1\}} \; [d \equiv n^j \bmod 24]. \tag{9.39b}$$

Condition (9.38) can not be omitted in Theorem 9.8. Indeed $n = 1729 = 7 \cdot 13 \cdot 19$ does satisfy (9.37), but does not satisfy (9.39b). Before we prove Theorem 9.8, we shall illustrate how it can be used to test the primality of integers n, $24 < n < 24^2$. After the proof we shall discuss generalizations of Theorem 9.8, that yield efficient primality tests for larger integers.

Step 1   Compute $gcd(6,n)$. If this gcd is not 1, we have an explicit factor of n.

Step 2   Check (9.37). If n does not satisfy (9.37), n can not be a prime by (9.23).

Step 3   Check (9.38). If n is prime, the probability that a random $1 < a < n$ satisfies (9.38) is 1/2. So in two tries one can expect to find an integer a satisfying (9.38). If no such integer a exists, n is not prime. More can be said about this step. Assuming the Extended Riemann Hypothesis one can prove that (9.38) has a solution a, $1 < a < 2(\log n)^2$, if n is prime. See also [Per86].

After these three steps we know from Theorem 9.8 that (9.39b) must hold.

Step 4   Check whether $n^1 \underline{\bmod} \; 24$ is greater than 1 and divides n. If so, n is not a prime.

If n "survives" these four tests, we may conclude that n is prime. Indeed any proper divisor d of n is congruent to 1 or n mod 24 by (9.39b). Since $n < 24^2$ we may assume that $d < 24$ (otherwise consider n/d instead of d). It follows that d is equal to the residue of 1 or n mod 24. The possibility $d = n \underline{\bmod} \; 24$, $d > 1$, is ruled out by step 4. It follows that $d = 1 \underline{\bmod} \; 24 = 1$. So n is prime.

In the proof of Theorem 9.8 we shall need the following necessary and sufficient condition for two integers n_1 and n_2, with $gcd(n_i,6) = 1$, $i = 1,2$, to be congruent to each other mod 24.

Lemma 9.9 Let n_1 and n_2 be two integers with $gcd(n_i,6) = 1$, $i = 1,2$. Then $n_1 \equiv n_2$ mod 24 iff

$$(i/n_1) = (i/n_2) \text{ for } i = -1, 2 \text{ and } 3.$$

Proof: There are eight integers $1 \le n \le 24$ with $gcd(n,6) = 1$, namely 1, 5, 7, 11, 13, 17, 19 and 23. It follows from (9.20) and (9.21) that for any n with $gcd(n,6) = 1$

$$(3/n) = 1 \text{ iff } n \equiv 1, 11, 13 \text{ or } 23 \text{ mod } 24.$$

From (9.21) and (9.22) it similarly follows that

$$(2/n) = 1 \text{ iff } n \equiv 1, 7, 17 \text{ or } 23 \text{ mod } 24.$$

$(-1/n) = 1$ iff $n \equiv 1, 5, 13$ or 17 mod 24.

So each of the integers 1, 5, 7, 11, 13, 17, 19 and 23 mod 24 is uniquely determined by the values of (i/n), $i = -1, 2, 3$ (see also Figure 9.7).

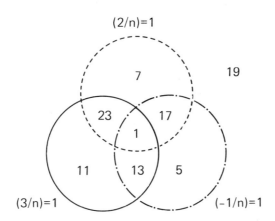

Figure 9.7: The intersection patterns of the integers n mod 24, $gcd(n, 6) = 1$,
with $(3/n) = 1$, $(2/n) = 1$ resp. $(-1/n) = 1$.

□

Proof of Theorem 9.8:

Since $gcd(n, 6) = 1$ implies that $n^2 \equiv 1$ mod 24, (9.39b) is a direct consequence of (9.39a). Next note that it suffices to prove (9.39a) for prime divisors d of n only. Write $n - 1 = u \cdot 2^k$ and $d - 1 = v \cdot 2^l$, u, v odd and $k, l > 0$. We shall first prove that $l \geq k$ and then use Lemma 9.9 to show that either $d \equiv n^0$ mod 24 or $d \equiv n^1$ mod 24. Raise both sides in (9.38) to the power v and reduce the result mod d. Since $d \mid n$ and v is odd, one obtains

$$a^{u \cdot v \cdot 2^{k-1}} \equiv a^{v(n-1)/2} \equiv -1 \text{ mod } d. \tag{9.40}$$

By Theorem 9.5 (Fermat)

$$a^{u \cdot v \cdot 2^l} \equiv a^{u(d-1)} \equiv 1^u \equiv 1 \text{ mod } d. \tag{9.41}$$

It follows from (9.40) and (9.41) that

$$k - 1 < l \tag{9.42}$$

Now consider $i \in \{-1, 2, 3\}$. Since v is odd and $d \mid n$, we deduce from (9.37)

$$i^{u \cdot v \cdot 2^{k-1}} \equiv i^{v(n-1)/2} \equiv (i/n)^v \equiv (i/n) \text{ mod } d. \tag{9.43}$$

On the other hand because d is prime, we have by (9.23) that

$$i^{u \cdot v \cdot 2^{l-1}} \equiv i^{u(d-1)/2} \equiv (i/d)^u \equiv (i/d) \bmod d. \qquad (9.44)$$

It follows from (9.43) and (9.44) that for $i = -1, 2, 3$

$$(i/d) = (i/n)^{2^{l-k}}. \qquad (9.45)$$

Note that we have replaced the congruence relation in (9.45) by an equality sign. We can do this, because both hands have value -1 or 1. If $l = k$, (9.45) and Lemma 9.9 imply that $d \equiv n \equiv n^1 \bmod 24$. If $l > k$, the right hand side of (9.45) is equal to 1, which is $(i/1)$. So Lemma 9.9 yields $d \equiv 1 \equiv n^0 \bmod 24$. By (9.42) we do not have to consider $k > l$. $\qquad \qquad \square$

Crucial in the application of Theorem 9.8 is the fact that we could replace (9.39a) by (9.39b). Because of this only one test was necessary in step 4. The reason that (9.39a) could be replaced by (9.39b) is the fact that

$$gcd(n, 24) = 1 \qquad \Rightarrow \qquad n^2 \equiv 1 \bmod 24 \qquad (9.46)$$

Theorem 9.8 can only prove the primality of integers n with $24 < n < 24^2$. For larger values of n one needs generalizations of Theorem 9.8. As may be expected, the exponent in (9.46) will have to be increased in these generalizations. For instance

$$gcd(n, 65520) = 1 \qquad \Rightarrow \qquad n^{12} \equiv 1 \bmod 65520.$$

In order to test 100-digit numbers for primality, one uses

$$gcd(n, s) = 1 \qquad \Rightarrow \qquad n^{5040} \equiv 1 \bmod s, \qquad (9.47)$$

where s is the 53-digit number

$$2^6 \cdot 3^3 \cdot 5^2 \cdot 7^2 \cdot 11 \cdot 13 \cdot 17 \cdot 19 \cdot 31 \cdot 37 \cdot 41 \cdot 43 \cdot 61 \cdot 71 \cdot 73 \cdot 113 \cdot 127 \cdot$$

$$181 \cdot 211 \cdot 241 \cdot 281 \cdot 337 \cdot 421 \cdot 631 \cdot 1009 \cdot 2521.$$

Note that $\sqrt{n} < s$, if n has not more than 100 digits. The rough outline of the primality test of a 100-digit number is as follows.

Step 1   Compute $gcd(n, s)$. If this is not 1, one has an explicit factor of n.

Step 2   Go through a list of 67 congruence relations like (9.37), that have to hold, if n is a prime. If n fails any of these tests, n is composite (but one does not have a divisor).

Step 3   Check if $n^i \bmod s$ divides n for $0 < i < 5040$. If so, one has a divisor of n. Otherwise n is prime.

Algorithm 9.7: The Cohen and Lenstra primality test.

If n is composite, the algorithm above will sometimes yield a factor of n. The probability that this happens however, is very small. In most cases it will be in step 2 that the algorithm stops and in that step one does not obtain a factor of n. The algorithm above can be adapted to test larger integers for primality. The expected running time is

$$O((\ln n)^{c \ln\ln n}), \tag{9.48}$$

where c is some constant.

§ 9.4 The Rabin variant

In § 9.1 it was mentioned that no other cryptanalysis of RSA is known than the factorization of n. In [Rab79] Rabin proposes a variant of the RSA system, whose cryptanalysis can be proved to be equivalent to the factorization of n. In the RSA system each user U had to select an integer $1 < e_U < \phi(n_U)$, with $gcd(e_U, \phi(n_U)) = 1$. In this variant all users U take

$$e_U = 2. \tag{9.49}$$

Since $gcd(2, \phi(n_U)) = 2$, encryption is no longer a one-to-one mapping. Indeed if $c \equiv m^2 \bmod n_U$, with $n_U = p_U \cdot q_U$, the congruence relations $x^2 \equiv c \bmod p_U$ and $x^2 \equiv c \bmod q_U$ each will have the two solutions $\pm m \bmod p_U$, resp. $\pm m \bmod q_U$. By the Chinese Remainder Theorem (Theorem A.23)

$$x^2 \equiv c \bmod n_U \tag{9.50}$$

has 4 solutions mod n_U. As a consequence this variant by Rabin does not have the signature property. How does one decrypt a message in this system? One precomputes integers a and b satisfying

$$a \equiv 1 \bmod p_U, \qquad b \equiv 0 \bmod p_U,$$

$$a \equiv 0 \bmod q_U, \qquad b \equiv 1 \bmod q_U.$$

The solutions a and b can easily be found by Euclid's Algorithm (see § A.2). Next one has to solve $x^2 \equiv c \bmod p_U$ (and similarly $x^2 \equiv c \bmod q_U$). If $c = 0$ this solution is obvious, so let us assume that $c \not\equiv 0 \bmod p_U$. For notational reasons we omit the subscripts. There are two cases to consider.

Case: $p \equiv 3 \bmod 4$.
The solution of $x^2 \equiv c \bmod p$ is given by $\pm c^{(p+1)/4}$. Indeed

$$(\pm c^{(p+1)/4})^2 \equiv c^{(p+1)/2} \equiv c \cdot c^{(p-1)/2} \equiv c \bmod p,$$

since by Fermat's Theorem (Theorems 9.5 or A.17)

$$c^{(p-1)/2} \equiv m^{p-1} \equiv 1 \bmod p.$$

Case: $p \equiv 1 \bmod 4$.
A fast deterministic algorithm to solve this congruence relation does not exist. We follow [Rab79].
Let **QR** denote the set of quadratic residues modulo p and **NQR** the set of quadratic non-residues modulo p. Let r and s be the solutions of $x^2 \equiv c \bmod p$. So $r \equiv m \bmod p$ and $s \equiv -m \bmod p$. Then $r + u$ and $s + u$ are the solutions of $(x - u)^2 - c \equiv 0 \bmod p$. Since $r \not\equiv s \bmod p$, it follows

that the field element (see Theorem B.13) $(r + u)/(s + u)$, $u \in \{0, 1, \ldots, p - 1\}\backslash\{-s\}$, takes on each value in $\{0, 1, \ldots, p - 1\}\backslash\{1\}$ exactly once. So for half of the admissible values of u the element $(r + u)/(s + u)$ will be in $\mathbf{QR} \cup \{0\}$ and for the other half it will be in $\mathbf{NQR}$. In the first case $r + u$ and $s + u$ will either both be in $\mathbf{QR}$ or they will both be in $\mathbf{NQR}$ or $u = -r$. In the latter case exactly one of them will be in $\mathbf{QR}$ and the other will be in $\mathbf{NQR}$. Since the $(p - 1)/2$ QR's are zero of $x^{(p-1)/2} - 1$ and this polynomial has degree $(p-1)/2$, one has

$$x^{(p-1)/2} - 1 = \prod_{a \in \mathbf{QR}} (x - a).$$

It follows from the above discussion that for a randomly chosen $u \in \{0, 1, \ldots, p-1\}\backslash\{-s\}$

$$gcd((x - u)^2 - c, x(x^{(p-1)/2} - 1)) \tag{9.51}$$

will be a linear polynomial with probability $(p - 1)/2(p - 1) = 1/2$. This linear polynomial will be $x - u - r$, if $u + r$ is in $\mathbf{QR}$ and $u + s$ is in $\mathbf{NQR}$, and $x - u - s$, if $u + r$ is in $\mathbf{NQR}$ and $u + s$ is in $\mathbf{QR}$. Since u is known, one can compute r resp. s. If $u = -s$, expression $(x - u)^2 - c$ will reduce to $x^2 + 2sx$. So the gcd in (9.51) will contain a factor x. A similar thing happens when $u = -r$. In all other cases the gcd will either be 1 (if $u + r$ and $u + s$ are both in $\mathbf{NQR}$), or $(x - u)^2 - c$, (if $u + r$ and $u + s$ are both in $\mathbf{QR} \cup \{0\}$). The expected number of u's that one has to try in this algorithm before finding a solution of $x^2 \equiv c \mod p$ is the reciprocal of $1/2$, i.e. 2. For a discussion of other methods of taking square roots modulo a prime number, we refer the interested reader to [Per86].

Let u be one of the two solutions of $x^2 \equiv c \mod p_U$ and v one of the the two solutions of $x^2 \equiv c$ mod q_U. Then by the Chinese Remainder Theorem (Theorem A.23) the four solutions of (9.50) are given by

$$(\pm u) \cdot a + (\pm v) \cdot b \mod n_U. \tag{9.52}$$

How two users can distinguish the four solutions and agree on one, depends on the actual choice of n_U. If p_U and q_U are both congruent to 3 mod 4, one has by (9.22) that -1 is a NQR mod p_U and mod q_U. Substitution of $u = -u$ and/or $v = -v$, if necessary, yields without loss of generality that

$u \cdot a + v \cdot b$	=	QR mod p_U	and	$u \cdot a + v \cdot b$	=	QR mod q_U
$u \cdot a - v \cdot b$	=	QR mod p_U	and	$u \cdot a - v \cdot b$	=	NQR mod q_U
$-u \cdot a + v \cdot b$	=	NQR mod p_U	and	$-u \cdot a + v \cdot b$	=	QR mod q_U
$-u \cdot a - v \cdot b$	=	NQR mod p_U	and	$-u \cdot a - v \cdot b$	=	NQR mod q_U

Note that $(ua + vb / p_U q_U) = (-ua - vb / p_U q_U) = 1$ while $(ua - vb / p_U q_U) = (-ua + vb / p_U q_U) = -1$. So there is a unique solution m' with $(m'/n_U) = 1$ and $0 < m' < n_U/2$. The receiver can easily determine m' from c in the above way.

It remains to show that breaking Rabin's variant of RSA is equivalent to factoring n_U.

Theorem 9.10 Let $n = p \cdot q$, p and q prime. Let A be an algorithm that finds a solution of $x^2 \equiv c \mod n$ in $F(n)$ steps for every c which is the square modulo n of an integer. Then a probabilistic algorithm exists that factors n in an expected number of $2(F(n) + 2\log_2 n)$ steps.

Proof: Select a random $0 < m < n$ and solve $x^2 \equiv m^2$ mod n in $F(n)$ steps with algorithm A. Let k be the solution found by A. The following four possibilities each have probability 1/4:

1) $k \equiv m$ mod p and $k \equiv m$ mod q .
2) $k \equiv m$ mod p and $k \equiv -m$ mod q .
3) $k \equiv -m$ mod p and $k \equiv m$ mod q .
4) $k \equiv -m$ mod p and $k \equiv -m$ mod q .

In case 2) $gcd(k - m, n) = p$ and in case 3) $gcd(k - m, n) = q$. So the computation of $gcd(k - m, n)$ will yield the factorization of n with probability 1/2. This computation needs less than $2\log_2 n$ steps. So each choice of m involves $F(n) + 2\log_2 n$ steps. Since the probability of success is 1/2 one expects to need two tries. ▯

Problems

1. Consider the RSA system with $n = 383 \times 563$ (so $n = 215629$) and encryption key $e = 49$. So a plaintext m will be encrypted as $c = E(m)$, where

    $$E(m) = m^{49} \underline{\text{mod}} \ n \ .$$

    Proof that every ciphertext c satisfies $E^{10}(c) = c$.

    Give an easy way for a cryptanalist to recover the plaintext m from the ciphertext c.

2. Verify that the decryption algorithm of the RSA system also recovers the plaintext m from the ciphertext, when m has a factor in common with the modulus n.

3. Consider the RSA cryptosystem with modulus $n = p \times q$ and encryption exponent e. Proof that the number of plaintexts m, satisfying

    $$m^e \equiv m \text{ mod } n \ ,$$

    (in which case the message is not concealed), is given by

    $$\{ 1 + gcd \ (e-1, p-1) \} \cdot \{ 1 + gcd \ (e-1, q-1) \} \ .$$

4. Demonstrate the principle of the Solovay and Strassen primality test on the number $m = 15$. The number m has been made small in this problem to keep the calculations simple. So do not make use of numbers that "incidentally" have a factor in common with m.

5.   Consider the Rabin variant of the RSA system. So only the number n is public. Suppose that a
      message m, $1 < m < n$, has been send that has a non-trivial factor in common with n. How
      many possible plaintexts will the receiver find at the end of the decryption process?

6.   The Rabin variant of the RSA system is used as cryptosystem with $n = 11 \times 19$. Demonstrate
      the decryption algorithm of this system for the ciphertext $c = 157$. Do not make use of the fact
      that 11 and 19 are small prime numbers. Other properties of p and q may of course be used.

10 THE MCELIECE SYSTEM

In this chapter it is assumed that the reader is familiar with algebraic coding theory. A reader without this background can freely skip this chapter and continue with Chapter 11. From [Mac77] we recall the following facts about *Goppa codes*.

> With each irreducible polynomial of degree t over $GF(2^m)$ corresponds a binary, irreducible Goppa code of length $n = 2^m$, dimension $k \geq n - tm$ and minimum distance $d \geq 2t + 1$. A fast decoding algorithm with running time nt, exists [Pat75]. There are about $2^{mt}/t$ (see Corollary B.23) irreducible polynomials of degree t over $GF(2^m)$. So a random polynomial of degree t over $GF(2^m)$ will be irreducible with probability $1/t$. Since there is a fast algorithm for testing irreducibility (see [Ber68, Ch.6] or [Rab80]), one can find an irreducible polynomial of degree t over $GF(2^m)$, just like in Algorithm 9.5, by repeated guessing and testing.

Based on this theory McEliece [McE78] proposed the following system. Each user U chooses a suitable $n_U = 2^{m_U}$ and t_U. User U selects a random, irreducible polynomial $p_U(x)$ of degree t_U over $GF(2^{m_U})$ and makes a generator matrix G_U of the corresponding Goppa code. The size of G_U is $k_U \times n_U$. Next user U chooses a random, dense, $k_U \times k_U$ nonsingular matrix S_U and a random $n_U \times n_U$ permutation matrix P_U and computes

$$G^* = S_U \, G_U \, P_U . \tag{10.1}$$

U makes G_U^* and t_U public, but keeps G_U, S_U and P_U secret.

Encryption: Suppose that user A wants to send a message to user B. He represents his message as binary strings $\underline{m}$ of length k_B, and sends to B

$$\underline{c} = \underline{m}\, G_B^* + \underline{e}\,,\qquad\qquad(10.2)$$

where $\underline{e}$ is random vector (error pattern) of length n_B and weight $\le t_B$.

Decryption: Upon receiving $\underline{c}$, B computes with his secret permutation matrix P_B

$$\underline{c}\,P_B^{-1} = \underline{m}\, S_B\, G_B\, P_B\, P_B^{-1} + \underline{e}\,P_B^{-1} = (\underline{m}\, S_B)\,G_B + \underline{e}'\,,$$

where $\underline{e}' = \underline{e}\,P_B^{-1}$ also has weight $\le t_B$. With the decoding algorithm of the Goppa code receiver B can now retrieve $\underline{m}\, S_B$. Multiplying this on the right with S_B^{-1} (only known to B) reproduces the original message $\underline{m}$.

Public	: G_U^* and t_U of all users U.
Secret	: $p_U(x)$, S_U and P_U by user U.
Property	: $S_U^{-1} G_U^* P_U^{-1}$ is a generator matrix of the t_U-error correcting Goppa code, corresponding to the irreducible polynomial $p_U(x)$ of degree t_U.
Messages from A to B	: binary vector $\underline{m}$ of length k_B.
Encryption by A	: $\underline{c} = \underline{m}\, G_B^* + \underline{e}$, weight $(\underline{e}) \le t_B$.
Decryption by B	: 1) compute $\underline{c}\,P^{-1} = : \underline{c}'$,
	2) decode $\underline{c}'$ to find $\underline{m}' = \underline{m}\, S_B$,
	3) compute $\underline{m} = \underline{m}' S_B^{-1}$.

Table 10.1 The McEliece system.

The McEliece system is summarized in Table 10.1. The reason that an error pattern is added in (10.2), is of course to make it difficult for the cryptanalist to retrieve $\underline{m}$ from $\underline{c}$. We shall now discuss the security of this system by analysing four possible attacks on a specific example.

Example 10.1: $m = 10$, $n = 1024$, $t = 50$, $k \approx 1024 - 50 \cdot 10 = 524$.

Attack 1:   Guess S_B and P_B to calculate G_B from G_B^*. One can now follow the decryption scheme of user B to find the message $\underline{m}$. However the number of different matrices S_B or P_B is so astronomical, that the probability of success of this attack is smaller than the probability of correctly guessing the vector $\underline{m}$ itself.

Attack 2: Compare the received vector $\underline{c}$ with all codewords in the code generated by G_B^* to find the closest codeword. So one now knows $\underline{m} \, G^*$. Retrieve $\underline{m}$ by a Gaussian elimination process. This approach involves $2^k \approx 2^{524} \approx 10^{158}$ comparisons!

Attack 3: Find the closest codeword to $\underline{c}$ by syndrome decoding and proceed as in attack 2. The work load is proportional to the number of syndromes, which is $2^{n-k} \approx 2^{500} \approx 10^{151}$!

Attack 4: Select k random positions and hope that they are not in error, i.e. hope that $\underline{e}$ has zero coordinates on these k positions. If these k positions are independent, $\underline{m}$ can be found with a Gaussian elimination process. Otherwise the k columns in G_B^* corresponding to these k positions, will very likely have rank close to k. So the Gaussian elimination process will lead to only a few possibilities for $\underline{m}$. Only one of these possibilities will correspond to a codeword at distance $\leq t_B$ from $\underline{c}$. The probability that the k positions are correct is about $(1 - t/n)^k$. The Gaussian elimination process involves k^3 steps. So the expected workload is

$$k^3(1 - t/n)^{-k} \approx 2^{65} \approx 10^{19} \,.$$

So even this attack is not feasible.

The heuristics behind this scheme are not difficult to guess. Take a sufficiently long, binary, linear block code, that can correct a large number, say t, of errors. The decoding algorithm on the other hand has to be fast. Manipulate the generator matrix to such a degree, that the new generator matrix looks like a random $k \times n$, rank k matrix. From complexity theory we know [Ber78] that the general decoding problem of linear codes is *NP-complete*. We shall not explain what that name means exactly. We refer the interested reader to [Gar79]. Here it suffices to know that this characterization implies that no known algorithm can solve this problem in a running time, depending polynomially on the size of the input. Moreover if one were to find such an algorithm, it could be adapted to solve a whole class of equally hard problems. The existing decoding schemes of general, binary, linear codes (attacks 2, 3 and 4) are not feasible by the size of the parameters.

The encryption function maps binary k-tuples into binary n-tuples. This mapping is not a surjection. Indeed, the number of vectors of length 1024 at distance ≤ 50 to a codeword is

$$2^k \sum_{i=0}^{t} \binom{n}{i} \approx 2^{524} \sum_{i=0}^{50} \binom{1024}{i} \approx 2^{808.4} \,,$$

which is an ignorable fraction of the total number of 1024-length words. So property PK4 in (7.4) does not hold. Consequently the McEliece system does not have the signature property.

11 THE KNAPSACK SYSTEM

§ 11.1 The knapsack system

In [Mer78] Merkle and Hellman propose a public key cryptosystem that is based on the difficulty of solving the knapsack problem. Later other knapsack related cryptosystems have been suggested. See, for instance [ChR85].

Knapsack problem:

Let $a_1, a_2, \ldots, a_n$ be a sequence of n integers. Let also S be an integer. Question: does the equation

$$x_1 a_2 + x_2 a_2 + \cdots + x_n a_n = S \tag{11.1}$$

have a solution with $x_i \in \{0, 1\}$, $1 \le i \le n$?

For large n the knapsack problem is hard to solve. In fact it has been shown that the knapsack problem is NP-complete (see page 82 or [Gar79]). Finding a $\{0, 1\}$-solution to (11.1) is of course at least as difficult as just finding out whether a solution exists. On the other hand for some sequences $\{a_i\}_{i=1}^{n}$ it is easy to find a $\{0, 1\}$-solution of (11.1), resp. demonstrating that no such solution exists. For example, given the sequence $a_i = 2^{i-1}$, $1 \le i \le n$, equation (11.1) will have a solution iff $0 \le S \le 2^n - 1$. Also finding the solution is very easy in this case. A much more general class of sequences $\{a_i\}_{i=1}^{n}$ exists, for which (11.1) is easily solvable. This is the class of superincreasing sequences. The sequence $\{a_i\}_{i=1}^{n}$ is called *superincreasing*, if for all $1 \le k \le n$

$$\sum_{i=1}^{k-1} a_i < a_k. \tag{11.2}$$

Algorithm 11.1 solves the knapsack problem for superincreasing sequences and actually finds the solution $\{x_i\}_{i=1}^{n}$ for each S, for which (11.1) is solvable. The knapsack problem for superincreasing sequences is easy to solve. Indeed a possible solution $\{x_i\}_{i=1}^{n}$ satisfies

$$x_n = 1 \quad \text{iff} \quad S \geq a_n,$$

and recursively for $j = n - 1, \ n - 2, \ldots, 1$

$$x_j = 1 \quad \text{iff} \quad S - \sum_{i=j+1}^{n} x_i \cdot a_i \geq a_j.$$

If $S - \sum_{i=1}^{n} x_i \cdot a_i = 0$ one has found the solution to (11.1), otherwise (11.1) does not admit a solution.

$$
\begin{aligned}
&j := n; \\
&\text{while } j \geq 1 \ do \\
&\qquad \text{begin} \\
&\qquad\quad \text{if } S \geq a_j \text{ then begin } x_j := 1; \ S := S - a_j \text{ end} \\
&\qquad\qquad\qquad \text{else } x_j := 0; \\
&\qquad\quad j := j - 1 \\
&\qquad \text{end}; \\
&\quad \text{if } S = 0 \text{ then print} \quad \text{"This knapsack has a solution".} \\
&\qquad\quad\quad\ \text{else print} \quad \text{"This knapsack does not have a solution".}
\end{aligned}
$$

Algorithm 11.1: Solving the knapsack problem for superincreasing sequences.

Based on the apparent difficulty of solving the knapsack problem and the ease to solve this problem for superincreasing sequences, the *Knapsack cryptosystem* has been proposed [Mer78].

Each user U makes a superincreasing sequence $\{u_i\}_{i=1}^{n_U}$ of length n_U, and selects integers M_U and W_U, such that

$$M_U > \sum_{i=1}^{n_U} u_i, \tag{11.3}$$

$$1 < W_U < M_U \text{ and } gcd(W_U, M_U) = 1. \tag{11.4}$$

User U computes the numbers

$$u_i{}' = W_U \cdot u_i \ \underline{\text{mod}} \ M_U, \ 1 \leq i \leq n_U, \tag{11.5}$$

and makes the sequence $\{u_i{}'\}_{i=1}^{n_U}$ known as his public key. He also computes $W_U^{-1} \ \underline{\text{mod}} \ M_U$ with Euclid's algorithm (see § A.2), but keeps this number, together with M_U, W_U and the original superincreasing sequence $\{u_i\}_{i=1}^{n_U}$ secret.

Encryption: If user A wants to send a message to user B, he looks up the public encryption key $\{b_i'\}_{i=1}^{n_B}$ of B. User A represents his message by binary sequences $\{m_i\}_{i=1}^{n_B}$ and sends to B the cipher-text

$$C = \sum_{i=1}^{n_B} m_i \cdot b_i' \, . \tag{11.6}$$

Decryption: User B computes with his secret W_B^{-1} and M_B

$$W_B^{-1} \cdot C \equiv W_B^{-1} \cdot \sum_{i=1}^{n_B} m_i \cdot b_i' \equiv \sum_{i=1}^{n_B} m_i \cdot b_i \bmod M_B \, . \tag{11.7}$$

By (11.3) $\sum_{i=1}^{n_B} m_i \cdot b_i < M_B$, so we can rewrite (11.7) as follows

$$\sum_{i=1}^{n_B} m_i \cdot b_i = (W_B^{-1} \cdot C \bmod M_B) \, . \tag{11.8}$$

Since the sequence $\{b_i\}_{i=1}^{n_B}$ is superincreasing, user B can now apply Algorithm 11.1 to $S = W_B^{-1} \cdot C \bmod M_B$ to recover the message $\{m_i\}_{i=1}^{n_B}$.

The system is summarized in Table 11.2.

Public	: $\{u_i'\}_{i=1}^{n_U}$ of all users U.
Secret	: $\{u_i\}_{i=1}^{n_U}$, W_U^{-1} and M_U by user U.
Property	: $u_i' = W_U u_i \bmod M_U$, $1 \le i \le n_U$,
	$\{u_i\}_{i=1}^{n_U}$ superincreasing, $gcd(W_U, M_U) = 1$.
Message of A for B	: Binary sequence $\{m_i\}_{i=1}^{n_B}$.
Encryption by A	: $C = \sum_{i=1}^{n_B} m_i \cdot b_i'$.
Decryption by B	: Solve the knapsack problem for the superincreasing sequence $\{b_i\}_{i=1}^{n_B}$ and $S = W_B^{-1} \cdot C \bmod M_B$.

Table 11.2: The knapsack cryptosystem.

Although the knapsack cryptosystem does not have the signature property, it gained an enormous popularity. The main reason is the simplicity to implement it. In applications both encryption and decryption can take place at very high data rates.

Example 11.1: $n_B = 6$, $M_B = 56789$ and $W_B = 12345$.

Let $b_1' = 44434$, $b_2' = 19714$, $b_3' = 56639$, $b_4' = 31669$, $b_5' = 44927$, $b_6' = 36929$ be the public key. For this small value of n_B it already takes some effort to decrypt (i.e. solve (11.6) for) $C = 101077$.

However the secret knapsack is the superincreasing sequence $b_1 = 22$, $b_2 = 89$, $b_3 = 345$, $b_4 = 987$, $b_5 = 4567$ and $b_6 = 45678$, as one can easily check with $W_B^{-1} = 39750$. To recover the message $\{m_i\}_{i=1}^6$ we compute $W_B^{-1} \cdot C \mod M_B$ and obtain $S = 39750 \times 101077 \mod 56789 = 45789$. With this S, the secret knapsack $\{b_i\}_{i=1}^6$ and Algorithm 11.1, it is easy to find $m_6 = 1$, $m_5 = 0$, $m_4 = 0$, $m_3 = 0$, $m_2 = 1$ and $m_1 = 1$.

The authors in [Mer78] recommend the users to take $n_U \approx 100$, any sequence $\{u_i\}_{i=1}^{n_U}$ with

$$(2^i - 1) \cdot 2^{100} < u_i < 2^i \cdot 2^{100}, \qquad 1 \le i \le 100, \tag{11.9}$$

(it will automatically be superincreasing) and

$$2^{201} + 1 < M_U < 2^{202}. \tag{11.10}$$

Note that also (11.3) is satisfied.

It is also a good idea for each user U, to publish a permuted version $\{u'_{\pi(i)}\}_{i=1}^{n_U}$, of his public knapsack and not $\{u_i'\}_{i=1}^{n_U}$ itself. In this way a cryptanalist has no information about which element u_i' in the public knapsack came from the smallest u_1, etc.. The idea of multiplying a superincreasing sequence with a constant W_U modulo M_U is of course to obtain a knapsack that looks random. To increase this effect and thus to increase the security of the knapsack cryptosystem, [Mer78] advises to iterate this multiplication. So each user U also selects $M_U' > \sum_{i=1}^{n_U} u_i'$ and $1 < W_U' < M_U'$ with $gcd(W_U', M_U') = 1$ and computes $u_i'' = W_U' \cdot u_i' \mod M'$, $1 \le i \le n_U$. In this modified version user U will publish the sequence $\{u_i''\}_{i=1}^{n_U}$ as his public knapsack and not $\{u'\}_{i=1}^{n_U}$. That it makes sense to iterate this process of modulo-multiplication, is illustrated in Example 11.2.

Example 11.2: $n = 3$

$u_1 = 5$	multiply by 17	$u_1' = 38$	multiply by 3	$u_1'' = 25$
$u_2 = 10$	==========>	$u_2' = 29$	==========>	$u_2'' = 87$
$u_3 = 20$	modulo 47	$u_3' = 11$	modulo 89	$u_3'' = 33$

It is impossible to find integers W and M that map (u_1, u_2, u_3) directly into (u_1'', u_2'', u_3''). Indeed the congruence relations

$$5 \cdot W \equiv 25 \mod M ,$$

$$10 \cdot W \equiv 87 \mod M$$

imply that M divides $87 - 2 \cdot 25 = 37$. Since 37 is a prime, it follows that $M = 37$. It also follows that $W = 5$. These values of W and M however violate the third congruence relation

$$20 \cdot W \equiv 33 \bmod M.$$

This shows that an iteration of modulo-multiplications can not always be replaced by a single modulo-multiplication.

Example 11.2 also demonstrates something else. A cryptanalist does not have to guess the original integers W_U and M_U ($+W_U'$ and M_U' in the iterated case) to find a superincreasing sequence. He can also decrypt the ciphertext, if he is able to obtain another superincreasing sequence from $\{u_i'\}_{i=1}^{n_U}$ (resp. from $\{u_i''\}_{i=1}^{n_U}$). In Example 11.2 we can also see that the second iteration mapped the not-superincreasing knapsack $\{u_i'\}_{i=1}^{n_U}$ into the (after proper reordering) superincreasing sequence $\{u_i''\}_{i=1}^{n_U}$! This demonstrates two important things:

1) Iteration does not necessarily increase the security of the system.

2) It may be easier for the cryptanalist to map the public knapsack into another superincreasing sequence than the original.

Some critics of the knapsack cryptosystem did not trust the linearity of the system. Their intuition/experience told them that the knapsack cryptosysten was bound to be broken. In the next paragraph we shall describe how Shamir [Sha82] did break the single multiplication version of the system. In § 11.3 an outline of the much more general attack by Lagarias and Odlyzko [Lag83] will be given.

§ 11.2 The Shamir attack

Since the general knapsack problem is NP-complete, no known algorithm solves it in polynomial time. However, the property of NP-completeness has never been proved for the restriction of the knapsack problem to the subclass of knapsacks, obtained by a single modulo-multiplication of a superincreasing sequence. The *Shamir attack* [Sha82], that we shall now discuss, consists of an algorithm, that solves the above mentioned, restricted knapsack problem with very high probability in polynomial time.

Let us assume that the cryptanalist is applying Shamir's algorithm to the cryptosystem of user U. For notational reasons we shall omit the subscripts. So the cryptanalist knows the public knapsack $\{u_i'\}_{i=1}^{n}$, which is obtained by i) multiplying the elements u_i, $1 \le i \le n$, of a superincreasing sequence by W mod M, followed by ii) a permutation P of the elements in the knapsack. Generalizing (11.9) and (11.10) we shall assume that u_i, $1 \le i \le n$, is $N + i$ bits long and that M is $N + n + 2$ bits long. In (11.9) and (11.10) we have $N = n = 100$.

In the subsequent algorithm we are going to find a pair $\hat{W}$ and $\hat{M}$, $gcd(\hat{W}, \hat{M}) = 1$, such that the sequence $\{\hat{W} \cdot u_i' \underline{\bmod} \hat{M}\}_{i=1}^{n}$ is a permutation of a superincreasing sequence. Such a pair will be called a "valid pair".

Because of the permutation P, exactly one of the $\binom{n}{4}$ different four-tuples $(u_{i_1}', u_{i_2}', u_{i_3}', u_{i_4}')$ will be equal to

$$(W \cdot u_1 \underline{\bmod} M \,, \ W \cdot u_2 \underline{\bmod} M \,, \ W \cdot u_3 \underline{\bmod} M \,, \ W \cdot u_4 \underline{\bmod} M) \,. \tag{11.11}$$

For each of the $\begin{bmatrix} n \\ 4 \end{bmatrix}$ possible choices of $(u_{i_1}', u_{i_2}', u_{i_3}', u_{i_4}')$ the subsequent steps will either not lead to a pair $\hat{W}$ and $\hat{M}$ at all, or lead to a pair $\hat{W}$ and $\hat{M}$, that is not valid (because the sequence $\{\hat{W}^{-1} \cdot u_i' \bmod \hat{M}\}_{i=1}^n$ is not a permutation of a superincreasing sequence), or lead to a valid pair $\hat{W}$ and $\hat{M}$. The right choice, i.e. the one corresponding to (11.11) will lead to a valid pair $\hat{W}$ and $\hat{M}$ with very high probability.

Without loss of generality we shall assume that (u_1', u_2', u_3', u_4') is the 4-tuple corresponding to (11.11). But the reader has to keep in mind that the workload in steps 1, 2, 3 and 4 has to be multiplied by $\begin{bmatrix} n \\ 4 \end{bmatrix}$.

Step 1:

Consider $\hat{W}$ as a variable for a moment. It is not possible to make a drawing of the mapping $\hat{W} \to \hat{W} \cdot u_1' \underline{\bmod} \hat{M}$, $1 < \hat{W} < \hat{M}$, simply because $\hat{M}$ is not known. Replacing $\hat{W}$ by $\hat{W}/\hat{M}$ and $\hat{M}$ by $\hat{M}/\hat{M} = 1$ solves this problem, as we shall see.

Let $V = W^{-1}/M$. So $0 < V < 1$. In Figure 11.3 an impression is given of the sawtooth function $f_1 : v \to v \cdot u_1' \bmod 1$, which maps $[0,1]$ into $[0,1)$. In reality the slope of f_1 will be much steeper. The zeros of f_1 will be correspondingly much closer to each other. More precisely the distance between two consecutive zeros of f_1 is equal to $1/u_1'$, which by the assumption on M satisfies

$$1/u_1' > 1/\hat{M} > 1/2^{N+n+2} \,. \tag{11.12}$$

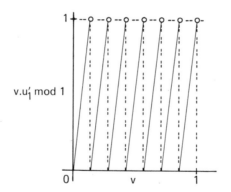

Figure 11.3 The mapping $f_1 : v \to v \cdot u_1' \underline{\bmod} 1$

It follows from our assumptions on $\{u_i\}_{i=1}^n$ and M, that

$$V \cdot u_1' \underline{\bmod} 1 = u_1/M < 2^{N+1}/2^{N+n+1} = 1/2^n \,. \tag{11.13}$$

So V will lie very close to the right of a zero of the sawtooth function f_1. More precisely, since the slope of the function f_1 is u_1', it follows from (11.13) that

$$0 < V - (\text{zero of } f_1, \text{ closest to } V) < (1/2^n)/u_1' . \qquad (11.14)$$

Hence the distance of V to the closest zero of f_1 is much smaller than the distance between two consecutive zeros of f_1. Assuming that W is randomly chosen with respect to M, $gcd(W,M) = 1$, one has with high probability that for $1 \le i \le 4$,

$$2^{N+n+2}/c < u_1' < 2^{N+n+2} , \qquad (11.15)$$

for some small constant c, say $c = 10$. It follows from (11.14) and (11.15) that

$$0 < V - (\text{zero of } f_1, \text{ closest to } V) < c/2^{N+2n+2} . \qquad (11.16)$$

In exactly the same way it can be shown, that V also lies very close to a zero of $f_i : v \to v \cdot u_i' \underline{\bmod} 1, i = 2,3,4$. More precisely

$$0 < V - (\text{zero of } f_i, \text{ closest to } V) < c \cdot 2^{i-1}/2^{N+2n+2} . \qquad (11.17)$$

By the triangle inequality

$$|(\text{zero of } f_i, \text{ closest to } V) - (\text{zero of } f_1, \text{ closest to } V)| \le$$

$$9 \cdot c/2^{N+2n+2} , \qquad i = 2,3,4 . \qquad (11.18)$$

Assuming that the numbers u_i, $1 \le i \le 4$, are independently selected, one may assume that also the numbers u_i', $1 \le i \le 4$, are independent. It follows that the probability that each of the functions f_i, $2 \le i \le 4$, has a zero at distance at most the right hand side of (11.18) to either side of a particular zero of f_1, is less than or equal to

$$2^{N+n+2} \cdot \left[\frac{2 \cdot 9 \cdot c/2^{N+2n+2}}{1/2^{N+n+2}} \right]^3 = 2^5 \cdot 3^6 \cdot c^3 \cdot 2^{N-2n} . \qquad (11.19)$$

The denominator in (11.19) is the lower bound for the distance between two consecutive zeros of f_1, as given by (11.12). The term outside the brackets in the left hand side of (11.19) is the upper bound for the number of zeros of f_i, $2 \le i \le 4$, on [0,1], given by the same (11.12).

For $N \approx n \approx 100$ the probability in (11.19) is very small. On the other hand this coincidence is exactly what happens around V. If one has other values of n and N, one may have to consider more than the four elements (u_1', u_2', u_3', u_4') in the knapsack to make the right hand side in (11.19) sufficiently small. So with very high probability there will be no other accumulation point of zeros of the functions f_i, $1 \le i \le 4$, than V. Let t_i/u_i' be the zero of f_i closest to V, $1 \le i \le 4$. Then it follows from the reasoning above, that

$$1 \le t_i < u_i' , \qquad 1 \le i \le 4 , \qquad (11.20)$$

$$\left| \frac{t_i}{u_i'} - \frac{t_1}{u_1'} \right| < 9 \cdot c/2^{N+2n+2} , \qquad 2 \le i \le 4 . \qquad (11.21)$$

The inequalities (11.20) and (11.21) give rise to an integer, linear programming problem in the variables t_i, $1 \leq i \leq 4$, that can be solved by the so called L^3-algorithm [Len82] in a computing time, that is polynomial in n (and N).

Step 2:

We conclude from the above that V lies in an interval $[a, a + \varepsilon)$, where a is the maximum of the zeros of f_i, $1 \leq i \leq 4$, and where ε is equal to the right hand side of (11.17) with $i = 4$. We also conclude that the four sawtooth functions f_i, $1 \leq i \leq 4$, are continuous on $[a, a + \varepsilon)$.

In this small interval other f_i's can have a zero. But V will still lie in the interval $[a^*, a + \varepsilon)$, where $a^* =$ the largest zero of the functions f_i in the interval $[a, a + \varepsilon)$, $1 \leq i \leq n$. By the size of ε no function f_i can have two zeros in $[a, a + \varepsilon)$. We conclude that V lies in the interval $[a^*, a + \varepsilon)$ and that each f_i, $1 \leq i \leq n$, is continuous on this interval.

Step 3: The n functions f_i, $1 \leq i \leq n$, will intersect in at most $\binom{n}{2}$ points in the interval $[a^*, a + \varepsilon)$. This leads to at most $\binom{n}{2} + 1$ subintervals of $[a^*, a + \varepsilon)$, on which the functions f_i have no further intersections. V will lie in one of these at most $\binom{n}{2} + 1$ intervals. On each of these intervals the functions f_i, $1 \leq i \leq n$, will have a constant vertical ordering. See Figure 11.4, where this situation has been depicted for only three functions.

This ordering determines the permutation P, that was applied to the sequence $\{ W \cdot u_i \bmod M \}_{i=1}^n$.

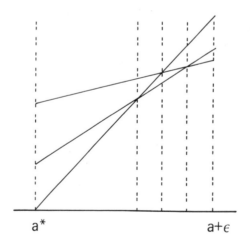

$$a^* \qquad\qquad a + \epsilon$$

Figure 11.4 An intersection pattern of three functions f_i on $[a^*, a + \varepsilon)$.

Step 4:

For each of the at most $\binom{n}{2} + 1$ subintervals I of $[a^*, a + \varepsilon)$, found in Step 3, we know the corresponding permutation P applied to the original knapsack. So we can write down the conditions on V, that are implied by the superincreasingness of the sequence $\{ V \cdot u'_{P^{-1}(i)} \bmod 1 \}_{i=1}^n$. These inequalities will define a subinterval I' of I, where they will all hold. At least one of the subintervals I', say $\hat{I}$, will be nonempty, because V lies in it. It follows that for any rational number $\hat{V} = \hat{W}/\hat{M}$ in $\hat{I}$

$$\{\hat{W} \cdot u'_{P^{-1}(i)} \bmod \hat{M}\}_{i=1}^{n} \text{ is a superincreasing sequence}. \tag{11.22}$$

With this superincreasing sequence $\{\hat{W} \cdot u'_{P^{-1}(i)} \bmod \hat{M}\}_{i=1}^{n}$ it is now easy to solve the knapsack problem.

In [Adl83] and [Bri85] the interested reader can find some information on how to break the iterated version of the Merkle-Hellman scheme.

11.3 The Lagarias and Odlyzko attack

This attack of the knapsack cryptosystem is much more fundamental than the Shamir attack. Up to now we have always assumed that the original $\{u_i\}_{i=1}^{n}$ sequence is superincreasing. This is however not essential for a knapsack-based cryptosystem. It only makes the decryption easy, because of Algorithm 11.1. Essential is only, that the mapping $\{m_i\}_{i=1}^{n} \to C$, in (11.6), is one-to-one.

Since the general knapsack problem is *NP*-complete, no known algorithm solves it in polynomial time. Still it is quite possible that polynomial-time algorithms do exist, which solve with a certain probability any knapsack problem in a large subclass of knapsack problems.

In this paragraph we shall often use the vector notation $\underline{u} = (u_1, u_2, \ldots, u_n)$ for a knapsack $\{u_i\}_{i=1}^{n}$. Before we give an outline of the Lagarias and Odlyzko attack [Lag83], we define a few new notions.

Definition 11.3: The *density* $d(\underline{u})$ of a knapsack $\underline{u}$ is defined by

$$d(\underline{u}) = \frac{n}{\log_2(\max_i u_i)}. \tag{11.23}$$

The density $d(\underline{u})$ is a measure for the information rate of the knapsack system. Indeed the numerator is the number of message bits that are stored in the sum C of the knapsack (see (11.6). The denominator is a good approximation of the average nummer of bits in the binary representation of C. For instance with $u_i = 2^{i-1}$, $1 \le i \le n$, $d(\underline{u}) = n/(n-1) \approx 1$, as it should be.

The smaller the density $d(\underline{u}')$ of a public knapsack $\underline{u}'$, the more likely the Lagarias and Odlyzko attack will break the system. This may sound like a heavy restriction, but it is not. First of all, if the density of the public knapsack is high, the knapsack can hardly hide the trapdoor (see [Lag84]). Secondly from a knapsack with high density one can obtain a knapsack with low density, simply by (repeated) multiplication with a constant W modulo a larger constant M, where $gcd(W, M) = 1$.

Definition 11.4: An *integer lattice* L is a subgroup of $(\mathbb{Z}^n, +)$, spanned by n vectors $\underline{v}_i$, $1 \le i \le n$, that are independent over $\mathbb{R}$. So

$$L = \left\{ \sum_{i=1}^{n} \alpha_i \underline{v}_i \,\middle|\, \alpha_i \in \mathbb{Z} , \ 1 \le i \le n \right\}. \tag{11.24}$$

The n independent vectors $\underline{v}_i$, $1 \le i \le n$, form a *basis* for the lattice L. The ordered basis $\underline{v}_i$,

$1 \leq i \leq n$, will be denoted by $[\underline{v}_1, \underline{v}_2, \ldots, \underline{v}_n]$.

The *Gram-Schmidt* process applied to the basis $[\underline{v}_1, \underline{v}_2, \ldots, \underline{v}_n]$, described by

$$\underline{v}_1^* = \underline{v}_1 \, ,$$

$$\underline{v}_2^* = \underline{v}_2 - \mu_{1,2} \underline{v}_1^* \, ,$$

$$\ldots$$

$$\underline{v}_n^* = \underline{v}_n - \mu_{1,n} \underline{v}_1^* - \mu_{2,n} \underline{v}_2^* - \cdots - \mu_{n-1,n} \underline{v}_{n-1}^* \, ,$$

where

$$\mu_{i,j} = \frac{(v_j, v_i^*)}{(v_i^*, v_i^*)} \, , \qquad 1 \leq i \leq j \leq n \, ,$$

yields the orthogonal basis $[\underline{v}_1^*, \underline{v}_2^*, \ldots, \underline{v}_n^*]$. The vectors $\underline{v}_i^*$, $1 \leq i \leq n$, in general will no longer have integer coordinates. Let $\| \underline{x} \|$, $\underline{x} = (x_1, x_2, \ldots, x_n)$, denote the standard Euclidean norm $\left[\sum_{i=1}^{n} x_i^2 \right]^{1/2}$.

Definition 11.5: A basis $[\underline{v}_1, \underline{v}_2, \ldots, \underline{v}_n]$ of an integer lattice L is called y-reduced, $1/4 \leq y < 1$, if in the terminology from above

i) $\| \underline{v}_i^* + \mu_{i-1,i} \underline{v}_{i-1}^* \|^2 \geq y \cdot \| \underline{v}_{i-1}^* \|^2 \, , \qquad 2 \leq i \leq n \, .$ \hfill (11.25)

ii) $| \mu_{i,j} | \leq 1/2 \, , \qquad 1 \leq i < j \leq n \, .$ \hfill (11.26)

An alternative definition of a y-reduced basis can be given as follows. Let V_k be the k-dimensional linear subspace of $I\!\!R^n$, spanned by $\underline{v}_i$, $1 \leq i \leq k$, or equivalently by $\underline{v}_i^*$, $1 \leq i \leq k$.

Let $V_k^\perp$ be the orthogonal complement of V_k. Define $\underline{v}_j^{(k)}$, $k+1 \leq j \leq n$, as the projection of $\underline{v}_j$ onto $V_k^\perp$. In particular $\underline{v}_{k+1}^{(k)}$ is $\underline{v}_{k+1}^*$. Then (11.25) and (11.26) are equivalent with

$$\| \underline{v}_i^{(i-2)} \|^2 \geq y \cdot \| \underline{v}_{i-1}^{(i-2)} \|^2 = y \| \underline{v}_{i-1}^* \|^2 \, , \qquad 2 \leq i \leq n \, ,$$ \hfill (11.25')

resp.

$$\| \underline{v}_j^{(i)} - \underline{v}_j^{(i-1)} \| \leq \frac{1}{2} \| \underline{v}_i^{(i-1)} \| \, , \qquad 1 \leq i < j \leq n \, .$$ \hfill (11.26')

In the sequel y will always be 3/4. The L_3-algorithm [Len82], which was already mentioned in the previous paragraph, is very effective in producing a y-reduced basis.

Theorem 11.6: Let $[\underline{v}_1, \underline{v}_2, \ldots, \underline{v}_n]$ be a basis of a lattice L, such that $\| \underline{v}_i \|^2 \leq B$, $1 \leq i \leq n$, for some constant B. The L^3-algorithm produces a reduced basis $[\underline{w}_1, \underline{w}_2, \ldots, \underline{w}_n]$ for L in

$$O(n^6 (\log B)^3) \text{ bit operations} \, ,$$ \hfill (11.27)

For our purpose also the following theorem [Len82, Prop.1.11] is very important.

Theorem 11.7: Let $[\underline{w}_1, \underline{w}_2, \ldots, \underline{w}_n]$ be a reduced basis of the lattice L. Then

$$\| \underline{w}_1 \|^2 \le 2^{n-1} \cdot \min_{\underline{x} \in L \setminus \{0\}} \| \underline{x} \|^2 . \tag{11.28}$$

As a matter of fact in [Len82, Prop.1.12] it is proved that no vector in a reduced basis can be very long. We can now present the so-called *SV*-algorithm (shortest vector algorithm), as described in [Lag83]. It is intended to find the solution of (11.1).

SV-algorithm:

Step 1:
Take the vectors

$$\underline{u}_1 = (1,0,...,0,-a_1) ,$$

$$\underline{u}_2 = (0,1,...,0,-a_2) ,$$

$$... \tag{11.29}$$

$$\underline{u}_n = (0,0,...,1,-a_n) ,$$

$$\underline{u}_{n+1} = (0,0,...,0,S) .$$

Together they form a basis $[\underline{u}_1, \underline{u}_2, \ldots, \underline{u}_{n+1}]$ for the $(n+1)$-dimensional lattice $L = L(\underline{u}, S)$.

Let $\underline{x} = (x_1, x_2, \ldots, x_n)$ be the $\{0,1\}$-solution of (11.1). It follows that $(x_1, x_2, \ldots, x_n, 0)$ is a vector in L of (very short) length $\le \sqrt{n}$.

Step 2:
Find a reduced basis $[\underline{w}_1, \underline{w}_2, \ldots, \underline{w}_{n+1}]$ of L with the L^3-algorithm [Len82].

Step 3:
Check if one of the $n+1$ "short" vectors $\underline{w}_i = (w_{i,1}, w_{i,2}, \ldots, w_{i,n+1})$, $1 \le i \le n+1$, has its first n coordinates equal to 0 or some constant α. For such a vector $\underline{w}$, check if $x_j = w_{i,j}/\alpha$, $1 \le j \le n$, is a solution of (11.1). If so, STOP. Otherwise continue with step 4.

Step 4:
Repeat steps 1, 2 and 3 with S replaced by $S' = \sum_{i=1}^{n} a_i - S$. A solution $(x_1', x_2', \ldots, x_n')$ of this knapsack problem will yield the solution $(1-x_1, 1-x_2, \ldots, 1-x_n)$ of the original knapsack.

The computing time of steps 1 and 3 is ignorable. So the running time of this algorithm essentially equals twice the running time of the L^3-algorithm, as given by (11.27). There is in no way a guarantee, that the *SV*-algorithm will find a solution of the knapsack problem. However the authors of [Lag83] give the following analysis of the *SV*-algorithm.

Theorem 11.8: Let $B \geq 2^{(1+\beta)n^2}$ for some constant $\beta > 0$. Then the number of knapsacks $\underline{a} = (a_1, a_2, \ldots, a_n)$ satisfying

i) $1 \leq a_i \leq B$, $1 \leq i \leq n$,

and

ii)   the SV-algorithm will find a solution of (11.1) for all S, for which (11.1) has a $\{0,1\}$-solution

is given by

$$B^n(1 - \varepsilon(B)) , \tag{11.30}$$

where

$$|\varepsilon(B)| \leq \frac{D(1+\beta)}{B^{c(\beta) - 3\frac{\ln n}{n}}} \tag{11.31}$$

for some constant D and $c(\beta) = 1 - (1+\beta)^{-1} > 0$.

Theorem 11.8 states that for any $\beta > 0$ and n sufficiently large one can solve the knapsack problem for almost all knapsacks $\underline{a} = (a_1, a_2, \ldots, a_n)$ with density

$$d(\underline{a}) \leq \frac{n}{\log_2 B} < \frac{1}{(1+\beta)n} . \tag{11.32}$$

With some additional work inequality (11.32) can be weakened to

$$d(\underline{a}) < (1-\varepsilon) \left[\log_2 \frac{4}{3} \right]^{-1} \frac{1}{n} , \tag{11.33}$$

for any fixed $\varepsilon > 0$. Inequality (11.33) is probably not best possible, but in [Lag83] it is conjectured that the right hand side of the best possible inequality on $d(\underline{a})$ will still tend to zero as n goes to infinity.

Problems

1.   Solve the knapsack problem if the elements are given by 333, 41, 4, 172, 19, 3, 80 and 11 and if the total size of the knapsack equals 227.

2.   Solve the knapsack problem if the elements are given by 31, 32, 46, 51, 63, 72 and 88 and if the total size of the knapsack equals 227.

3.   A knapsack cryptosystem has the numbers 381, 424, 2313, 2527, 2535, 3832, 3879 and 4169 as public key. They are obtained by multiplying the elements of a superincreasing sequence by $w = 4673$ and reducing the result modulo 5011 (so $w^{-1} = 934$). Decrypt the message 1656.

4. Let $p_1, p_2, \cdots, p_n$ be a sequence of different prime numbers and let P be their product. The numbers a_i, $1 \le i \le n$, are defined by $a_i = P / p_i$. Let $S = \Sigma_{1 \le i \le n} x_i a_i$, where the elements x_i are either 0 or 1. Give a simple algorithm to recover the number x_i, $1 \le i \le n$, from S.
    Give an upperbound of the density of this knapsack.

12 THRESHOLD SCHEMES

In this chapter we shall not introduce a new cryptosystem, but we shall discuss a related topic. We start with an example [Liu68].

Example 12.1:

"Eleven scientists are working on a secret project. They wish to lock up the documents in a cabinet so that the cabinet can be opened if and only if six or more of the scientists are present. What is the smallest number of locks needed? What is the smallest number of keys to the locks each scientist must carry?"

Clearly for each 5-tuple of scientists there is at least one lock, that can not be opened by them. Also each of the six remaining scientists does have a key of that lock. More than one such lock per 5-tuple is not needed. So $\begin{bmatrix} 11 \\ 5 \end{bmatrix} = 462$ locks are needed and each scientist carries $\begin{bmatrix} 11-1 \\ 5 \end{bmatrix} = 252$ keys.

The solution above is of course not very practical. We shall consider here a more general situation.

Definition 12.2: Let D be some secret data (e.g. a safe combination). D is taken from a finite set U. Let n participants each have some partial information D_i about D, $D_i \in U$, $1 \le i \le n$. Then $\{D_i\}_{i=1}^{n}$ is called an (k,n)-*threshold scheme*, if

T_1:   Knowledge of k or more different D_i's makes D easily computable.

T_2:   Knowledge of $k-1$ or less D_i's leaves the secret D completely undetermined (i.e. all possible values in U are equally likely candidates for D).

In [Sha79] one can find the following construction of (k,n)-threshold schemes.

Let q be a prime power, $q > n$, and let α be a primitive element of $GF(q)$ (see §B.4). Let $D \in GF(q)$ be the secret data. Consider the polynomial

$$f(x) = D + a_1 x + \cdots + a_{k-1} x^{k-1} , \tag{12.1}$$

where a_i, $1 \le i \le k-1$, are independently selected, random elements in $GF(q)$. The n participants are given the pairs (i, D_i), where

$$D_i = f(\alpha^i) , \qquad 1 \le i \le n . \tag{12.2}$$

To check that the pairs (i, D_i), $1 \le i \le n$, form a (k,n) threshold scheme, we distinguish two cases.

Ad T_1:   Suppose that $D_{i_1}, D_{i_2}, \ldots, D_{i_k}$, $i_1 < i_2 < \cdots < i_k$ are known. With the *Lagrange interpolation formula*, it is quite easy to determine $f(x)$. Indeed

$$f(x) = \sum_{j=1}^{k} D_j \prod_{l=1, l \ne j}^{k} \frac{(x - \alpha^{i_l})}{(\alpha^{i_j} - \alpha^{i_l})} , \tag{12.3}$$

since it has degree $k-1$ and has the right value for $x = \alpha^{i_u}$, $1 \le u \le k$.

From (12.1) it follows that the value of D is given by $f(0)$.

Ad T_2:   Suppose that $D_{i_1}, D_{i_2}, \ldots, D_{i_l}$, $i_1 < i_2 < \cdots < i_l$, are known for some $l < k$. It follows from (12.1) and (12.3) that there are exactly q^{k-l-1} polynomials $f(x)$ with $f(\alpha^{i_u}) = D_{i_u}$, $1 \le u \le l$, and with prescribed value $f(0)$.

In [McE81] a variant of the construction above is proposed, that can handle the situation that some of the participants provide false values of D_i. Some participants may want to do this to prevent others from getting access to the secret data. As in Chapter 10, the theory of error-correcting codes is needed. We refer the reader, who is not familiar with this theory, to [Mac77,Chapter 11].

Theorem 12.3: Let q be a prime power, $q - 1 = n$. Let $\alpha_1, \alpha_2, \ldots, \alpha_n$ be a list of the different non-zero elements in $GF(q)$. Let $D \in GF(q)$ be a (secret) element in $GF(q)$. Define $f(x)$ as in (12.1). The n participants are given the pairs (i, D_i), where $D_i = f(\alpha_i)$, $1 \le i \le n$. Suppose that $D_{i_1}^*, D_{i_2}^*, \ldots, D_{i_s}^*$, $i_1 < i_2 < \cdots < i_s$, are given, with $D_j^* \ne D_j$ for exactly t values of $j \in \{i_1, i_2, \ldots, i_s\}$. Then D can be efficiently recovered, if $s \ge k + 2t$.

Proof: The vector $(D_1, D_2, \ldots, D_n)$ is a codeword in the Reed-Solomon code of length n, dimension k and minimum distance $d = n - k + 1$. This code is capable of correcting t errors and r erasures if

$$2t + r \leq d - 1 = n - k .$$

(12.4)

Moreover, efficient algorithms exists [Ber68, § 10.4],[Sug76], that can correct these errors and erasures. The $n - s$ unknown D_i's can be viewed as $n - s$ erasures. The t values of j for which $D_j^* \neq D_j$ correspond to t errors. So the above mentioned decoding algorithms will yield the vector $(D_1, D_2, \ldots, D_n)$, provided that $2t + (n - s) \leq n - k$, i.e. $s \geq k + 2t$.

The secret D can be found by computing $-\sum_{i=1}^{n} D_i$. Indeed, writing $a_0 = D$,

$$-\sum_{i=1}^{n} D_i = -\sum_{i=1}^{n} f(\alpha_i) = -\sum_{i=1}^{n} \sum_{j=0}^{k-1} a_j \, \alpha_i^j =$$

$$= -(q-1)a_0 - \sum_{j=1}^{k} a_j \sum_{\substack{\alpha \in GF(q) \\ \alpha \neq 0}} \alpha^j = D + \sum_{j=1}^{k-1} a_j \cdot 0 = D .$$

☐

Remark 12.4: By shortening the Reed-Solomon code, if necessary, one can extend the above construction to all values $n \leq q - 1$.

Remark 12.5: For $t = 0$ the above Theorem reduces to the [Sha79]-construction.

Remark 12.6: If more than $[(n - k)/2]$ participants give false information, the secret data D can be made inaccesible.

Problems

1. Set up a Shamir (3,5)-threshold sheme for the secret 15 in GF(17).

    Show how participants 1,2 and 3 can recover the secret.

    Show that for participants 1 and 2 together each element in GF(17) is an equally likely candidate for the secret.

13 OTHER DIRECTIONS

The best way to get an impression of the developments in cryptology that are not discussed in this textbook, is to look at the proceedings of the conferences held on cryptology. There are two yearly conferences on cryptology. Since 1981 one takes place in Santa Barbara and is called Crypto '81, Crypto '82, etc. The other one has been held in various cities in Europe since 1982 and is called Eurocrypt 82, etc. The IEEE Symposia on Information Theory that are held every one and a half year also always have several sessions on cryptology. Many articles on cryptology have appeared in the IEEE Transactions on Information Theory and are also to be expected in the forthcoming new Journal on Cryptology. Cryptologia is a quarterly journal on cryptology since 1978. It contains many interesting articles of a historical nature, but also new results do appear in it.

Here we briefly want to mention three such research areas. We shall give references to relevant literature.

Randomized Encryption. The key idea here is to have a cryptosystem in which many ciphertexts do correspond to one plaintext. A random mechanism can determine which ciphertext to choose. The McEliece system, see Chapter 10, is an example of randomized encryption. There the choice of the error pattern e was done in a random way.
Of course, by using a randomized encryption one hopes to achieve greater cryptographic security. The following example makes this clear.

Example 13.1: Let S be a message source generating symbols from $\{0,1\}$ with an independent, but identical distribution given by $p(0) = 1/4$ and $p(1) = 3/4$. This symbol 0 will be encrypted into one of the strings 00, 01, 10 and 11, say ab, and the symbol 1 will be randomly encrypted into one of the other three strings, each with the same probability. The cryptanalist can not decrypt the ciphertext, because each of the four possibilities of ab leads to an equally likely plaintext. This system is unconditionally secure.

The prize that one pays for this higher security is in the data expansion that takes place. The ciphertext is longer than the plaintext. Also, because encryption is now no longer a one-to-one mapping, one will never have the signature property PK4 (see (7.4)). But the advantages indicated above may easily outweigh these disadvantages. The interested reader is invited to read [GMi84] or [Riv83].

Protocols. A cryptographic protocol is a sequence of instructions that the various parties in a communication system have to follow in order to establish a connection and to exchange information. Two such protocols can be found in Problem 7.1. See also [DoY81].
Very interesting work has been done by D. Chaum and people around him (see [Cha85] and [ChE87]). They study so-called *credential mechanisms* that enable individuals to do business with organizations without transferring unnecessary information. It prevents these organizations from exchanging information about such an individual. Typically an individual will use a different pseudonym with each organization. So the organization still gets the information that it needs, like "this is a valid check", without revealing unnecessary things like the name and address of the individual.

Zero knowledge proofs. These are techniques (see [GMR85]) to convince somebody else that one has certain knowledge, without revealing even one bit of information of that knowledge. We shall give an example of how this can be done.

Example 13.2: Prover A presents two integers u and n to verifier B and claims that he also knows an integer x such that $x^2 \equiv u \bmod n$. By techniques described in § 9.4 such an integer x can easily be found if n is a prime, but in general finding such a solution seems to be as hard as factoring n.
How does A convince B that he indeed knows x. To this end the following dialogue between A and B takes place.

Step 1: A – picks a random y,
             – computes $v \equiv y^2 \bmod n$
             – and presents v to B.

Step 2a:	B asks A the solution of $y^2 \equiv v \bmod n$.	Step 2b: B asks A the solution of $y^2 \equiv uv \bmod n$.
Step 3a:	A gives y to B.	Step 3b: A gives $z = xy$ to B.
Step 4a:	B checks $y^2 \equiv v \bmod n$.	Step 4b: B checks $z^2 \equiv uv \bmod n$.

In this scheme it is up to B to opt for the a-steps or the b-steps. Very likely B will randomly choose between these two possibilities.

Note that A can give y resp. xy to B in Step 3 if he knows x. If A does not know x, he will with probability 1/2 get Question 2b and will not be able to give the correct value of z in Step 3b.

Note on the other hand that A only reveals y or xy, where y is a randomly chosen number. So B does not obtain any information about the secret x. The above procedure is repeated k-times, where k is sufficiently large to convince B of the correctness of A's claim of knowing x.

In [Blu86] the author gives a method that for every mathematical theorem T enables a prover A to convince a verifier B that A indeed has the proof of theorem T, while not giving away any information about that proof.

APPENDIX A   ELEMENTARY NUMBER THEORY

§ A.1 Introduction

Let $I\!N$ denote the set of natural numbers, $\mathbb{Z}$ the set of integers and $I\!R$ the set of real numbers.

An integer d *divides* an integer n, if $n = k \cdot d$ for some $k \in \mathbb{Z}$. We shall denote this by $d \mid n$. If such an integer k does not exist, d does not divide n. This is denoted by $d \nmid n$.

An integer p, $p > 1$, is said to be *prime*, if 1 and p are its only positive divisors. With $p_1 = 2, p_2 = 3, p_3 = 5, \ldots$ we introduce a natural numbering of the set of prime numbers.

Theorem A.1 (Euclid, 300 B.C.)

There are infinitely many prime numbers.

Proof. Suppose the contrary. Let $p_1, p_2, \ldots, p_k$ be the set of all primes. Then we obtain a contradiction, by observing that the integer $\left[\prod_{i=1}^{k} p_i \right] + 1$ is not divisible by any of the primes p_i, $1 \le i \le k$. []

Between two consecutive primes there can be an arbitrary large gap of non-prime numbers. For example the $n-1$ elements in the sequence $n! + 2, n! + 3, \ldots, n! + n$ are divisible by 2 resp., 3, $\ldots, n$. So none of them is prime.

Definition A.2 The function $\pi : I\!R^+ \to I\!N$ is defined by

$$\pi(x) = |\{p \text{ prime } | p \leq x\}| \ . \tag{A.1}$$

In words: $\pi(x)$ counts the number of primes less than or equal to x. The next theorem, which we shall not prove, tells us something about the relative frequency of the prime numbers in $I\!N$.

Theorem A.3 (The *Prime Number Theorem*, [Har45, p. 9])

$$\lim_{x \to \infty} \frac{\pi(x) \cdot \ln(x)}{x} = 1 \ . \tag{A.2}$$

Definition A.4 The *greatest common divisor* of two integers a and b is the uniquely determined, positive integer d, satisfying

$$d \mid a \quad \text{and} \quad d \mid b \tag{A.3}$$

and

$$\forall_{f \in I\!N} \ [(f \mid a \text{ and } f \mid b) \Rightarrow f \mid d] \ . \tag{A.4}$$

It is denoted by $gcd(a,b)$ or simply (a,b).

Similarly the *least common multiple* of two integers a and b is the uniquely determined, positive integer m, satisfying

$$a \mid m \quad \text{and} \quad b \mid m \tag{A.5}$$

$$\forall_{f \in I\!N} \ [(a \mid f \text{ and } b \mid f) \Rightarrow m \mid f] \ . \tag{A.6}$$

It is denoted by $lcm[a,b]$ or simply $[a,b]$.

For the existence of *ggd* we introduce the set $U = \{xa + yb \mid x \in \mathbb{Z}, y \in \mathbb{Z}, xa + yb > 0\}$. Let $m := \min U$. Clearly if $f \mid a$ and $f \mid b$ then $f \mid m$. So (A.4) is satisfied by m. Now write $a = qm + r$, $0 \leq r < m$. If $r \neq 0$, then $r \in U$, contradicting our assumption on m. So $r = 0$ i.e. $m \mid a$. Similarly $m \mid b$. So m satisfies (A.3) too. The uniqueness of $gcd(a,b)$ follows from (A.3) and (A.4). Indeed, if d and d' both satisfy (A.3) and (A.4), it follows that $d \mid d'$ and $d' \mid d$. So $d = d'$. In a similar way the uniqueness of $lcm[a,b]$ can be proved.

Alternative definitions of $gcd(a,b)$ and $lcm[a,b]$ are

$$gcd(a,b) = \max\{d > 0 \mid d \mid a \wedge d \mid b\} \ ,$$

$$lcm[a,b] = \min\{m > 0 \mid a \mid m \wedge b \mid m\} \ .$$

Theorem A.5 Let a and b be in $I\!N$. Then there exist integers u and v, such that

$$ua + vb = gcd(a,b) \ . \tag{A.7}$$

Proof: Let d be the smallest positive integer in the set

$$D = \{xa + yb \mid x \in \mathbb{Z}, y \in \mathbb{Z}\} . \tag{A.8}$$

Clearly $gcd(a,b)$ divides every element in D. In particular $gcd(a,b) \mid d$. So it suffices to show that $d \mid gcd(a,b)$. Obviously $0 < d \le a$. Let q and r be defined by $a = qd + r$, $0 \le r < d$. Then for some x and y in $\mathbb{Z}$

$$r = a - qd = a - q(xa + yb) = (1 - qx)a - (qy)b .$$

So $r \in D$. It follows from our assumption on d that $r = 0$, i.e. $d \mid a$. In a similar way one can show that $d \mid b$. Hence, by (A.4), $d \mid gcd(a,b)$. □

We can now give a formal proof of the following "obvious" lemma.

Lemma A.6 Let $d \mid ab$ and $gcd(d,a) = 1$. Then $d \mid b$.

Proof: Since $gcd(d,a) = 1$, Theorem A.5 implies that $xd + ya = 1$, for some integers x and y. So $xdb + yab = b$. Since $d \mid ab$, it follows that d divides $(xdb + yab) = b$. □

Corollary A.7 Let p be prime and $p \mid a_1 a_2 \cdots a_k$, where $a_i \in \mathbb{Z}$, $1 \le i \le k$. Then p divides at least one of the factors a_i, $1 \le i \le k$.

Proof: Use Lemma A.6 and induction on k. □

With an induction argument the following theorem can now easily be proved.

Theorem A.8 (Fundamental Theorem of Number Theory)
Any positive integer has a unique factorization of the form

$$\Pi_i \, p_i^{e_i} , \qquad e_i \in \mathbb{N} . \tag{A.9}$$

Let $a = \Pi_i \, p_i^{e_i}$, $e_i \in \mathbb{N}$, and $b = \Pi_i \, p_i^{f_i}$, $f_i \in \mathbb{N}$. Then one easily checks that

$$gcd(a,b) = \Pi_i \, p_i^{\min\{e_i, f_i\}} , \tag{A.10}$$

$$lcm[a,b] = \Pi_i \, p_i^{\max\{e_i, f_i\}} , \tag{A.11}$$

$$gcd(a,b) \cdot lcm[a,b] = ab . \tag{A.12}$$

§ A.2 Euclid's Algorithm

Let a and b be two positive integers with $b \ge a$. Clearly any divisor of a and b is a divisor of a and $b - a$ and vice versa. So $gcd(a,b) = gcd(a, b - a)$. Writing $b = q \cdot a + r$, $0 \le r < a$, one has for the same reason that $gcd(a,b) = gcd(r,a)$. If $r = 0$, we may conclude that $gcd(a,b) = a$ otherwise we continue in the same way with a and r. So we write $a = q' \cdot r + r'$, $0 \le r' < r$, etc. This algorithm is an extremely fast way of computing the gcd of two integers and it is known as *Euclid's*

Algorithm.

If one also wants to find the coefficients u and v satisfying (A.7), this algorithm can be adapted as described below. Note that by leaving out the lines involving the integers u and v, this (extended) algorithm reduces to simple version above.

$$\text{(initialize)} \quad s_0 = b \; ; \quad s_1 = a \; ;$$
$$u_0 = 0; \quad u_1 = 1;$$
$$v_0 = 1; \quad v_1 = 0;$$
$$n = 1;$$

while $s_n > 0$ do

$$\text{begin } n = n + 1; \quad q_n = \lfloor s_{n-2}/s_{n-1} \rfloor \; ;$$
$$s_n = s_{n-2} - q_n s_{n-1};$$
$$\text{(so } s_n \text{ is the remainder of } s_{n-2} \text{ divided by } s_{n-1})$$
$$u_n = q_n u_{n-1} + u_{n-2};$$
$$v_n = q_n v_{n-1} + v_{n-2}$$

end;

$$u = (-1)^n u_{n-1}; \qquad v = (-1)^{n-1} v_{n-1}; \tag{A.13}$$

$$gcd(a, b) = s_{n-1} . \tag{A.14}$$

Proof: First observe that the elements s_n, $n \geq 1$, form a strictly decreasing sequence of non-negative integers. So the algorithm will terminate after at most b iterations. Later in this paragraph we shall analyze how fast Euclid's Algorithm really is.

From the recurrence relation $s_k = s_{k-2} - q_k s_{k-1}$ in the algorithm it follows that

$$gcd(a, b) = gcd(s_0, s_1) = gcd(s_1, s_2) = \cdots =$$

$$= gcd(s_{n-1}, s_n) = gcd(s_{n-1}, 0) = s_{n-2} .$$

This proves (A.14).

Now we shall prove that for all k, $o \leq k \leq n$,

$$(-1)^{k-1} u_k a + (-1)^k v_k b = s_k . \tag{A.15}$$

For $k = 0$ and $k = 1$ (A.15) is true by the initialization values that u_0, u_1, v_0 and v_1 got in the algorithm. We now proceed by induction. It follows from the recurrence relations in the algorithm and from the induction hypothesis, that

$$s_k = s_{k-2} - q_k s_{k-1} = \{(-1)^{k-3} u_{k-2} a + (-1)^{k-2} v_{k-2} b\} - q_k \{(-1)^{k-2} u_{k-1} a + (-1)^{k-1} v_{k-1} b\} =$$

$$= (-1)^{k-1} \{u_{k-2} + q_k u_{k-1}\} a + (-1)^k \{v_{k-2} + q_k v_{k-1}\} b = (-1)^{k-1} u_k a + (-1)^k v_k b .$$

This proves (A.15) for all k, $0 \leq k \leq n$. Substitution of $k = n - 1$ in (A.15) yields

$$(-1)^n u_{n-1} a + (-1)^{n-1} v_{n-1} b = s_{n-1} .$$ (A.16)

Comparison of (A.16) with (A.7) proves (A.13). □

Of course there is no need to keep all the previously calculated values of s_k, u_k and v_k stored in the program. Only the last two of each will suffice. So the required memory space is a constant. Before we derive the number of operations involved in the algorithm, we study its behaviour on certain starting pairs a and b.

Let $\{F_n\}_{n=0}^{\infty}$ be the sequence of *Fibonacci numbers* numbers defined recursively by $F_0 = 0$, $F_1 = 1$ and $F_n = F_{n-1} + F_{n-2}$, $n \geq 2$.

It is easy to see that Euclid's Algorithm applied to F_{n-1} and F_n involves exactly $n-2$ iterations (all q_i's are equal to 1). Now

$$F_n = \frac{1}{\sqrt{5}} \cdot \left[\frac{1+\sqrt{5}}{2} \right]^n - \frac{1}{\sqrt{5}} \cdot \left[\frac{1-\sqrt{5}}{2} \right]^n ,$$ (A.17)

as one can easily prove with an induction argument or with the theory of generating functions. Since F_n and $((1+\sqrt{5})/2)^n$ both satisfy the same recurrence relation it is straightforward to prove with an induction argument that

$$\left[\frac{1+\sqrt{5}}{2} \right]^{n-2} < F_n < \left[\frac{1+\sqrt{5}}{2} \right]^{n-1} .$$ (A.18)

So instead of n iterations one can say that the computation of $gcd(F_{n-1}, F_n)$ takes $\lfloor \log_f (F_n) \rfloor$ iterations, where $f = (1 + \sqrt{5})/2$.

Theorem A.9 Let a and b be positive integers, $b \geq a$, and let $f = (1+\sqrt{5})/2$. Then the number of iterations, that Euclid's Algorithm will need to compute $gcd(a,b)$, is at most

$$\log_f (b) .$$ (A.19)

Proof: Note that f is the largest root of $x^2 = x + 1$.

The statement of the theorem can easily be verified for small values of b. We proceed with induction on b. Write $b = qa + r$, $0 \leq r < a$. We distinguish two cases.

Case 1: $a \leq b/f$.

By the induction hypothesis Euclid's Algorithm computes $gcd(a,r)$ using at most $\log_f (a)$ iterations. So for the computation of $gcd(b,a)$ at most

$$1 + \log_f (a) = \log_f (f \cdot a) \leq \log_f (b)$$

iterations are needed.

Case 2: $b/f < a \leq b$.

It follows that $q = 1$, $b = a + r$ and $0 \leq r = b - a < b(1 - f^{-1}) = b/f^2$. Writing $a = q'r + r'$, $0 \leq r' < r$, it follows from the induction hypothesis that Euclid's Algorithm needs at most $\log_f(r)$ iterations to compute $gcd(r, r')$. So for $gcd(a, b)$ at most

$$2 + \log_f(r) = \log_f(f^2 \cdot r) < \log_f(b)$$

iterations are needed.   []

It follows from the example of the Fibonnaci numbers that the constant f in (A.19) can not be replaced by a larger value. In previous chapters we have often replaced the bound $\log_f(b)$ by the weaker bound $2\log_2(b)$.

§ A.3 Congruences, Fermat, Euler, Chinese Remainder Theorem

Let a and b be two integers. Then a is said to be *congruent* to b *modulo* m, if $b - a$ is divisible by m. This is denoted by

$$a \equiv b \bmod m .$$ (A.20)

Definition A.10 A set of m integers $a_1, a_2, \ldots, a_m$ is called a *complete residue system* modulo m, if each integer j is congruent to (exactly) one of the elements a_i, $1 \leq i \leq m$, modulo m.

Clearly the m integers a_i, $1 \leq i \leq m$, form a complete residue system modulo m if and only if for each pair $1 \leq i, j \leq m$ one has that

$$a_i \equiv a_j \bmod m \Rightarrow i = j .$$ (A.21)

The congruence relation $\equiv \bmod m$ defines an equivalence relation (see Definition B.8) on $\mathbb{Z}$. A complete residue system is just a set of representatives of the m equivalence classes.

Lemma A.11 If $ka \equiv kb \bmod m$ and $gcd(k, m) = d$, then $a \equiv b \bmod m/d$.

Proof: Write $k = k'd$ and $m = m'd$ with $gcd(k', m') = 1$. It follows from $ka - kb = xm$, $x \in \mathbb{Z}$, that $k'(a - b) = xm'$. Since $gcd(m', k') = 1$, it follows from Lemma A.6 that $m' \mid (a - b)$, i.e. $a \equiv b \bmod m'$.   []

Lemma A.12 Let $a_1, a_2, \ldots, a_m$ be a complete residue system modulo m and let $gcd(k, m) = 1$. Then $ka_1, ka_2, \ldots, ka_m$ is also a complete residue system modulo m.

Proof: We use criterion (A.21). By Lemma A.11 $ka_i \equiv ka_j \bmod m$ implies that $a_i \equiv a_j \bmod m$. This in turn implies that $i = j$.

□

Often we shall only be interested in representatives of those residue classes modulo m, whose elements have gcd 1 with m. The number of these classes is denoted by the following function.

Definition A.13 (Euler's Totient Function $\phi(m)$).

$$\phi(m) = |\{0 \leq r < m \mid gcd(r,m) = 1\}| . \tag{A.22}$$

Theorem A.14 For all positive integers n

$$\sum_{d \mid n} \phi(d) = n . \tag{A.23}$$

Proof: Let $d \mid n$. By writing $r = id$ one sees immediately that the number of elements r, $0 \leq r < n$, with $gcd(r,n) = d$ is equal to the number of integers i with $0 \leq i < n/d$ and $gcd(i,n/d) = 1$. So this number is $\phi(n/d)$. On the other hand, $gcd(r,n)$ divides n for each integer r, $0 \leq r < n$. It follows that $\sum_{d \mid n} \phi(n/d) = n$, which is equivalent to (A.23). □

A set of $\phi(m)$ integers $r_1, r_2, \ldots, r_{\phi(m)}$ is called a *reduced residue system* modulo m if each integer j, $gcd(j,m) = 1$, is congruent to (exactly) one of the elements r_i, $1 \leq i \leq \phi(m)$.

Analogously to Lemma A.12 one has the following lemma.

Lemma A.15 Let $r_1, r_2, \ldots, r_{\phi(m)}$ be a reduced residue system modulo m and let $gcd(a,m) = 1$. Then $ar_1, ar_2, \ldots, ar_{\phi(m)}$ is also a reduced residue system modulo m.

With Lemma A.15 one can easily prove that the classes in a reduced residue system form a multiplicative group (see § B.1).

Theorem A.16 (Euler)

$$gcd(a,m) = 1 \Rightarrow a^{\phi(m)} \equiv 1 \bmod m . \tag{A.24}$$

Proof: Let $r_1, r_2, \ldots, r_{\phi(m)}$ be a reduced residue system modulo m. By Lemma A.15

$$\prod_{i=1}^{\phi(m)} r_i \equiv \prod_{i=1}^{\phi(m)} (ar_i) \equiv a^{\phi(m)} \cdot \prod_{i=1}^{\phi(m)} r_i \bmod m .$$

Since $gcd(\prod_{i=1}^{\phi(m)} r_i, m) = 1$, (A.24) follows from Lemma A.11. □

Let p be a prime number. Since every integer r, $1 \leq r < p$, has gcd 1 with p, it follows that

$\phi(p) = p - 1$ for p prime. This makes the next theorem a special case of Theorem A.16.

Theorem A.17 (Fermat)

Let p be a prime and $p \nmid a$. Then

$$a^{p-1} \equiv 1 \bmod p .$$

As we have just observed, $\phi(p) = p - 1$ for p prime. Because exactly one of every p consecutive integers is divisible by p, we have the following stronger result:

$$\phi(p^e) = p^e - p^{e-1} = p^{e-1}(p-1) = p^e(1 - p^{-1}) . \tag{A.25}$$

Definition A.18 A function $f : \mathbb{N} \to \mathbb{N}$ is said to be *multiplicative*, if for every pair of positive integers m and n

$$gcd(m, n) = 1 \;\Rightarrow\; f(mn) = f(m)f(n) .$$

Lemma A.19 Euler's Totient function $\phi(m)$ is multiplicative.

Proof : Let $gcd(m, n) = 1$ and let $a_1, a_2, \ldots, a_{\phi(m)}$ and $b_1, b_2, \ldots, b_{\phi(n)}$ be reduced residue systems modulo m resp. n. It suffices to show that the $\phi(m) \cdot \phi(n)$ integers $na_i + mb_j$, $1 \le i \le \phi(m)$ and $1 \le j \le \phi(n)$, form a reduced residue system modulo mn. It is quite easy to check that the integers $na_i + mb_j$, $1 \le i \le \phi(m)$ and $1 \le j \le \phi(n)$, are all different modulo mn and that they have gcd 1 with mn.

It remains to verify that any integer k, $gcd(k, mn) = 1$, is congruent to $na_i + mb_j \bmod mn$ for some $1 \le i \le \phi(m)$ and $1 \le j \le \phi(n)$.

From Lemma A.15 we know that integers i and j, $1 \le i \le \phi(m)$ and $1 \le j \le \phi(n)$, exist for which

$$k \equiv na_i \bmod m \quad \text{and} \quad k = mb_j \bmod n .$$

This implies that both m and n divide $k - na_i - mb_j$.

Since $gcd(m, n) = 1$, it follows from (A.6) and (A.12), that also mn divides $k - na_i - mb_j$.   []

Corollary A.20

$$\phi(m) = m \cdot \Pi_{p \mid m} (1 - p^{-1}) . \tag{A.26}$$

Proof: Combine (A.25) and Lemma A.19.   []

The simplest congruence relation, that one may have to solve, is the single, linear congruence relation

$$ax \equiv b \bmod m . \tag{A.27}$$

Theorem A.21 The linear congruence relation $ax \equiv b \bmod m$ has a solution x if and only if $gcd(a,m) \mid b$. In this case the number of different solutions modulo m is $gcd(a,m)$.

Proof: That $gcd(a,m) \mid b$ is a necessary condition for (A.27) to have a solution x is trivial. We shall now prove that $gcd(a,m) \mid b$ is also a sufficient condition for (A.27) to have solution. Let $d = gcd(a,m)$ and write $a = a' \cdot d$, $m = m' \cdot d$ and $b = b' \cdot d$. Then $gcd(a',m') = 1$. By Lemma A.12 the congruence relation $a'x \equiv b' \bmod m'$ has a unique solution x' modulo m'. Clearly a solution x of $ax \equiv b \bmod m$ satisfies $x \equiv x' \bmod m'$. So each solution x modulo m can be writen as $x' + im'$, $0 \le i \le d-1$. Write $a'x' = b' + um'$, $u \in \mathbb{Z}$. Then for each $0 \le i \le d-1$, $a(x'+im') = = da'x' + ida'm' = db' + udm' + ia'm = b + (u+ia')m$. Hence the numbers $x' + im'$, $0 \le i \le d-1$, are all the solutions modulo m of $ax \equiv b \bmod m$. □

The solution of $ax \equiv b \bmod m$, $gcd(a,m) = 1$, can easily be found with Euclid's Algorithm. Indeed from $ua + vm = 1$ (see (A.7)), it follows that $ua \equiv 1 \bmod m$. So the solution x is given by $bu \bmod m$. If $gcd(a,m) = 1$, one often writes a^{-1} for the unique element u satisfying $ua \equiv 1 \bmod m$.

Example A.22 To solve $14x \equiv 26 \bmod 34$, we first solve $7x' \equiv 13 \bmod 17$. With Euclid's Algorithm we find $5 \times 7 + (-2) \times 17 = 1$. So $7 \cdot 5 \equiv 1 \bmod 17$ and x' can be computed from $x' \equiv 7^{-1} \cdot 13 \equiv \equiv 5 \cdot 13 \equiv 14 \bmod 17$. By Theorem A.21, $14x \equiv 26 \bmod 34$ has the numbers 14 and 31 as solutions mod 34.

We shall now discuss the case that x has to satisfy several, linear congruence relations simultaneously, say $a_i x \equiv b_i \bmod m_i$ with $gcd(a_i,m_i) \mid b_i$ for $1 \le i \le k$. Dividing the i-th relation by $d_i = gcd(a_i,m_i)$, $1 \le i \le k$, one gets as before the congruence relation $a_i'x' \equiv b' \bmod m_i'$, with $gcd(a',m') = 1$. By the proof of Theorem A.21 a solution of this congruence relation is equivalent to a solution of one of the d_i congruence relations $a_i'x \equiv b_i' + jm_i' \bmod m_i$, $0 \le j \le d_i-1$. In view of this, we restrict our attention to the case that $gcd(a_i,m_i) = 1$ for all i, $1 \le i \le k$.

Theorem A.23 (The *Chinese Remainder Theorem*)

Let m_i, $1 \le i \le k$, be pairwise coprime integers. Let a_i, $1 \le i \le k$, be integers with $gcd(a_i,m_i) = 1$. Then the system of k simultaneous congruence relations

$$a_i x \equiv A_i \bmod m_i , \qquad 1 \le i \le k , \tag{A.28}$$

has a unique solution mod $m_1 m_2 \cdots m_k$ for all possible k-tuples of integers $A_1, A_2, \ldots, A_k$.

Proof: Suppose that x' and x'' both satisfy (A.28). Then $a_i(x'-x'') \equiv 0 \bmod m_i$, $1 \le i \le k$. By Lemma A.6 m_i divides $x'-x''$ for all $1 \le i \le k$. It follows that $x' \equiv x'' \bmod m_1 m_2 \cdots m_k$. So each of the $m_1 m_2 \cdots m_k$ possible k-tuples of right hand sides $A_1, A_2, \ldots, A_k$ corresponds with a unique integer solution x, $0 \le x < m_1 m_2 \cdots m_k$, of (A.28).   ▯

The proof above does not give an efficient algorithm to determine the solution of (A.28). Let $1 \le i \le k$ and let u_i be the unique solution of

$$a_i u_i \equiv 1 \bmod m_i \ , \tag{A.29}$$

$$a_j u_i \equiv 0 \bmod m_j \ , \qquad 1 \le j \le k \ , \ j \ne i \ . \tag{A.30}$$

With Euclid's Algorithm u_i is easy to determine. Indeed from (A.30) it follows that $u_i = r m^{(i)}$ for some $0 \le r < m_i$, where $m^{(i)} = \Pi_{j \ne i} \, m_j$. The value of r follows from (A.29). Indeed r is the solution of $a_i m^{(i)} r \equiv 1 \bmod m_i$. Hence

$$u_i = \{ (a_i m^{(i)})^{-1} \bmod m_i \} \, m^{(i)} \tag{A.31}$$

The numbers u_i, $1 \le i \le k$, can be stored using at most $k \log_2 m$ bits of memory space. The solution of (A.28) is now given by

$$x = u_1 A_1 + u_2 A_2 + \cdots + u_k A_k \ . \tag{A.32}$$

In Example 8.4 one can find a nice application of the Chinese Remainder Theorem.

For the computation of the numbers u_i, $1 \le i \le k$, by (A.31) we have to divide $m = \prod_{j=1}^{k} m_j$ by m_i (to find $m^{(i)}$) and we have to apply Euclid's Algorithm, which takes at most $\log_f (m)$ iterations for each i, $1 \le i \le k$. In this specific application of Euclid's Algorithm we are only interested in the $\{u_n\}_n$-sequence and not in the $\{v_n\}_n$-sequence.

The final computation of the solution x (by (A.32)) only involves k multiplications and $k - 1$ additions. Together this proves the following Theorem.

Theorem A.24 After a precomputation that only depends on the values of a_i and m_i, $1 \le i \le k$, and that uses at most

$$O(k \log_f (m)) \text{ operations}$$

and

$$O(k \log_2(m)) \text{ bits of memory space} \ ,$$

where $m = \prod_{j=1}^{k} m_j$, the solution of (A.28) can be determined in

$$2k \text{ operations} \ .$$

All these operations involve numbers of at most $\log_2(m)$ bits long.

§ A.4 Quadratic residues

Let p be an odd prime. The quadratic congruence relation $ax^2 + bx + c \equiv 0 \bmod p$ can be simplified by i) dividing the congruence relation by a and ii) the substitution $x \to x - b/2a$. In this way $ax^2 + bx + c \equiv 0 \bmod p$ reduces to a quadratic congruence relation of the type:

$$x^2 \equiv u \bmod p \ . \tag{A.33}$$

Definition A.25 Let p be an odd prime and u an integer such that $p \nmid u$. Then u is called a *quadratic residue* (**QR**), if (A.33) has an integer solution, and *quadratic non-residue* (**NQR**), if (A.33) does not have an integer solution.

Definition A.26 Let p be an odd prime and u an integer. The *Legendre symbol* (u/p) is defined by

$$(u/p) = \begin{cases} 1 \ , & \text{if } u \text{ is a quadratic residue mod } p \quad , \\ -1 \ , & \text{if } u \text{ is a quadratic non–residue mod } p \ , \\ 0 \ , & \text{if } p \mid u \end{cases}$$

If there is no confusion about the actual choice of the prime number p, one often writes $\chi(u)$ instead of (u/p).

Let $a^2 \equiv u \bmod p$. Then also $(p-a)^2 \equiv u \bmod p$. The polynomial $x^2 - u$ has at most two solutions in GF(p) (see Theorem B.17). It follows that the quadratic residues modulo p are given by the integers $i^2 \bmod p$, $1 \le i \le \dfrac{p-1}{2}$, or alternatively by the integers $(p-i)^2 \bmod p$, $1 \le i \le \dfrac{p-1}{2}$. We conclude that there are exactly $(p-1)/2$ **QR**'s and $(p-1)/2$ **NQR**'s. This proves the first of the following two theorems.

Theorem A.27 Let p be an odd prime. Then exactly $(p-1)/2$ of the integers $1, 2, \ldots, p-1$ are quadratic residue and $(p-1)/2$ are quadratic non-residue. In formula

$$\sum_{u=0}^{p-1} \chi(u) = 0 \ . \tag{A.34}$$

Theorem A.28 Let p be an odd prime. Then for all integers u and v

$$\chi(u) \cdot \chi(v) = \chi(uv) \ . \tag{A.35}$$

Proof: This theorem will be a trivial consequence of Theorem A.30. We shall present here a more elementary proof.

If $p \mid u$ or $p \mid v$ the assertion is trivial. The proof for $p \nmid u$ and $p \nmid v$ is split up in three cases.

Case 1: u and *v* are both **QR**.

Then $u \equiv a^2 \bmod p$ and $v \equiv b^2 \bmod p$, for some integers a and b. It follows that $uv \equiv (ab)^2 \bmod p$. So uv is **QR**.

Case 2: Exactly one of u and v is **QR**, say u is **QR** and v is **NQR**.

Suppose that also uv is **QR**. Then there exist integers a and b such that $u \equiv a^2 \bmod p$ and $uv \equiv b^2 \bmod p$. Since $a \not\equiv 0 \bmod p$, it follows that $v \equiv (b/a)^2 \bmod p$. A contradiction!

Case 3: u and v are both **NQR**.

From Lemma A.12 we know that iu, $i = 1, 2, \ldots, p-1$, runs through all non-zero elements. For the $(p-1)/2$ values of i for which i is **QR**, we have by case 2 that iu is **NQR**. So for the $(p-1)/2$ values of i for which i is **NQR**, it follows that iu is **QR**. So uv is **QR**. $\square$

Although the next theorem has never been used in this textbook, we do mention it, because it is often needed in related areas.

Theorem A.29 Let p be an odd prime. Then for every integer v

$$\sum_{u=0}^{p-1} \chi(u)\chi(u+v) = \begin{cases} p-1, & \text{if } p \mid v, \\ -1, & \text{if } p \nmid v. \end{cases} \tag{A.36}$$

Proof: If $p \mid v$, the statement is trivial. For $p \nmid v$ one has by (A.35) and (A.34) that

$$\sum_{u=0}^{p-1} \chi(u)\chi(u+v) = \sum_{u \neq 0} \chi(u)\chi(u+v) = \sum_{u \neq 0} \chi(u)\chi(u)\chi(1+v/u) =$$

$$= \sum_{u \neq 0} \chi(1+v/u) = \sum_{w \neq 1} \chi(w) = -1 + \sum_{w=0}^{p-1} \chi(w) = -1 . \qquad \square$$

Let u be **QR**, say $u \equiv a^2 \bmod p$. By Fermat's Theorem $u^{(p-1)/2} \equiv a^{p-1} \equiv 1 \bmod p$. So the $(p-1)/2$ **QR**'s are zero of the polynomial $x^{(p-1)/2} - 1$ over GF(p). Since a polynomial of degree $(p-1)/2$ over GF(p) has at most $(p-1)/2$ different zeros in GF(p) (see Theorem B.17), one has

$$x^{(p-1)/2} - 1 = \Pi_{u \text{ is } \mathbf{QR}} \, (x - u) . \tag{A.37}$$

It also follows that $u^{(p-1)/2} \neq 1$, if u is **NQR**. Since $(u^{(p-1)/2})^2 \equiv 1 \bmod p$ by Fermat's Theorem and since $y^2 \equiv 1 \bmod p$ has only 1 and -1 as zeros, it follows that $u^{(p-1)/2} \equiv -1 \bmod p$, if u is **NQR**. This proves the following theorem for all u with $p \nmid u$. For $p \mid u$ the theorem is trivially true.

Theorem A.30 Let p be an odd prime. Then for all integers u,

$$(u/p) \equiv u^{(p-1)/2} \bmod p . \tag{A.38}$$

Corollary A.31 Let p be an odd prime. Then

$$(-1/p) = \begin{cases} 1, & \text{if } p \equiv 1 \bmod 4, \\ -1, & \text{if } p \equiv 3 \bmod 4. \end{cases} \tag{A.39}$$

Proof: $(-1)^{(p-1)/2} = 1$ if and only if $p \equiv 1 \bmod 4$.

Another value of the Legendre symbol that we shall need later on is $(2/p)$.

Theorem A.32 Let p be an odd prime. Then

$$(2/p) = \begin{cases} 1, & \text{if } p \equiv \pm 1 \bmod 8, \\ -1, & \text{if } p \equiv \pm 3 \bmod 8. \end{cases} \tag{A.40}$$

Proof:

$$2^{(p-1)/2} \cdot \prod_{k=1}^{(p-1)/2} k \equiv \prod_{k=1}^{(p-1)/2} 2k \equiv \left[\prod_{k=1}^{\lfloor (p-1)/4 \rfloor} 2k \right] \cdot \left[\prod_{k=1+\lfloor (p-1)/4 \rfloor}^{(p-1)/2} 2k \right] \equiv$$

$$\equiv (-1)^{(p-1)/2 - \lfloor (p-1)/4 \rfloor} \cdot \prod_{k=1}^{\lfloor (p-1)/4 \rfloor} 2k \cdot \prod_{k=1+\lfloor (p-1)/4 \rfloor}^{(p-1)/2} (p - 2k) \equiv$$

$$(-1)^{(p-1)/2 - \lfloor (p-1)/4 \rfloor} \cdot \left[\prod_{k=1}^{(p-1)/2} k \right] \bmod p.$$

Dividing the above relation by $\prod_{k=1}^{(p-1)/2} k$ yields

$$2^{(p-1)/2} \equiv (-1)^{(p-1)/2 - \lfloor (p-1)/4 \rfloor} \bmod p.$$

The theorem now follows from (A.38). □

Before we describe a fast way to compute the Legendre symbol, we shall first generalize it to all odd integers.

Definition A.33 Let $m = \prod p_i^{e_i}$, $e_i \geq 0$, be an odd integer and let u an integer with $gcd(u, m) = 1$. Then the *Jacobi* symbol (u/m) is defined by

$$(u/m) = \prod_i (u/p_i)^{e_i}, \tag{A.41}$$

where (u/p_i) is the Legendre symbol.

Theorem A.34 Let m and n be odd integers. Then the following relations hold for the Jacobi symbol

$$(u/m) = ((u-m)/m), \tag{A.42}$$

$$(uv/m) = (u/m)(v/m) \; , \tag{A.43}$$

$$(u/mn) = (u/m)(u/n) \; , \tag{A.44}$$

$$(-1/m) = 1 \text{ iff } m \equiv 1 \bmod 4 \; , \tag{A.45}$$

$$(2/m) \;\; = 1 \text{ iff } m \equiv \pm 1 \bmod 8 \; . \tag{A.46}$$

Proof: Relations (A.42) and (A.43) hold for the Legendre symbol and by (A.41) also for the Jacobi symbol. Relation (A.44) is a direct consequence of (A.41). To see that relation (A.45) is a direct consequence of (A.41) and Corollary A.31, it suffices to observe that a product of an odd number of integers, each congruent to 3 mod 4, is also congruent to 3 mod 4 , while for an even number the product will be 1 mod 4. The proof of relation (A.46) goes analogously.   []

One more relation is needed to be able to compute (u/m) fast. We shall not give its proof, because the theory involved goes beyond the scope of this book. The interested reader is referred to [Har45, Thm. 99] or [Shp83, Thm. 7.2.1].

Theorem A.35 (*Quadratic Reciprocity Law* by Gauss)
Let m and n be odd coprime integers. Then

$$(m/n)(n/m) = (-1)^{(m-1)(n-1)/4} \tag{A.47}$$

With relations (A.42)-(A.47) one can evaluate (u/m) very fast. See Example 9.3.

§ A.5 Möbius inversion formula, the principle of inclusion and exclusion

Often in Discrete Mathematics a function f is defined in terms of of another function, say g. The question is, how g can be expressed in terms of f. With the theory of partially ordered sets and the (generalized) Möbius inversion formula, one can frequently solve this problem [Aig79, IV]. In this paragraph we shall discuss two important special cases, that follow from the theory, mentioned above, but which can also be proved directly.

Often we shall need an explicit factorization of an integer n. We do not longer want the strict ordering of the prime numbers, given by $p_1 = 2, p_2 = 3$, etc.. However different subscripts will still denote different prime numbers.

Definition A.36 Let $n = \prod_{i=1}^{k} p_i^{e_i}, e_i > 0, 1 \le i \le k$. Then the *Möbius function* $\mu(n)$ is defined by

$$\mu(n) = \begin{cases} 1 \; , & \text{if } n = 1 \; , \\ 0 \; , & \text{if } e_i \ge 2 \text{ for some } i \; , \; 1 \le i \le k \; , \\ (-1)^k \; , & \text{if } e_1 = e_2 = \cdots = e_k = 1 \; . \end{cases} \tag{A.48}$$

In other words $\mu(n)$ is the multiplicative function satisfying $\mu(1) = 1, \mu(p) = -1$ and $\mu(p^i) = 0, i \ge 2,$

for any prime p.

The Möbius function is defined in this peculiar way so that it has the following property.

Theorem A.37 Let n be a positive integer. Then

$$\sum_{d \mid n} \mu(d) = \begin{cases} 1, & \text{if } n = 1, \\ 0, & \text{if } n > 1. \end{cases} \tag{A.49}$$

Proof: For $n = 1$ the assertion is trivial. For $n > 1$ we write as above $n = \prod_{i=1}^{k} p_i^{e_i}$, $e_i > 0$, $1 \le i \le k$. Then $k > 0$ and thus

$$\sum_{d \mid n} \mu(d) = \sum_{d \mid p_1 p_2 \cdots p_k} \mu(d) = 1 + \sum_{l=1}^{k} \sum_{1 \le i_1 < i_2 < \cdots < i_l \le k} \mu(p_{i_1} p_{i_2} \cdots p_{i_l}) =$$

$$= \sum_{l=0}^{k} \binom{k}{l} (-1)^l = (1-1)^k = 0. \qquad \square$$

Theorem A.38 Let m and n be two positive integers such that $m \mid n$. Then

$$\sum_{\substack{d \\ m \mid d \mid n}} \mu(n/d) = \begin{cases} 1, & \text{if } m = n, \\ 0, & \text{otherwise}. \end{cases} \tag{A.50}$$

Proof: Let $n = n'm$. For each d, $m \mid d \mid n$, write $d = d'm$. Then $\sum_{d, m \mid d \mid n} \mu(n/d) = \sum_{d' \mid n'} \mu(n'/d')$, which by (A.49) is 1 for $n' = 1$, (i.e. $m = n$), and 0 for $n' > 1$. $\qquad \square$

Theorem A.39 (*Möbius Inversion Formula*)

Let f be a function defined on N. Define the function g on N by

$$g(n) = \sum_{d \mid n} f(d), \qquad n \in N. \tag{A.51}$$

Then for all $n \in N$

$$f(n) = \sum_{d \mid n} \mu(d) g(n/d) = \sum_{d \mid n} \mu(n/d) g(d). \tag{A.52}$$

Proof: By (A.51) and (A.50)

$$\sum_{d \mid n} \mu(n/d) g(d) = \sum_{d \mid n} \mu(n/d) \sum_{e \mid d} f(e) = \sum_{e \mid n} f(e) \sum_{\substack{d \\ e \mid d \mid n}} \mu(n/d) = f(n). \qquad \square$$

Corollary A.40 (*Multiplicative Möbius Inversion Formula*)

Let F be a function defined on $I\!N$. Let the function G be defined on $I\!N$ by

$$G(n) = \prod_{d \mid n} F(d) , \quad n \in I\!N .$$

Then for all $n \in I\!N$

$$F(n) = \prod_{d \mid n} G(n/d)^{\mu(d)} = \prod_{d \mid n} G(d)^{\mu(n/d)} .$$

Proof: Substitute $g(n) = \log(G(n))$ and $f(n) = \log(F(n))$ in Theorem A.39.   []

Example A.41 From Theorem A.14 we know that Euler's Totient Function as defined by (A.22) satisfies

$$\sum_{d \mid n} \phi(d) = n .$$

It follows from Theorem A.39 that for $n = \prod_{i=1}^{k} p_i^{e_i}$, $e_i > 0$, $1 \le i \le k$

$$\phi(n) = \sum_{d \mid n} \mu(d) \frac{n}{d} =$$

$$= \frac{n}{1} - \sum_{1 \le i \le k} \frac{n}{p_i} + \sum_{1 \le i < j \le k} \frac{n}{p_i p_j} - \cdots + (-1)^k \frac{n}{p_1 p_2 \cdots p_k} =$$

$$= n \left[1 - \frac{1}{p_1} \right] \left[1 - \frac{1}{p_2} \right] \cdots \left[1 - \frac{1}{p_k} \right] .$$

This proofs (A.26) in a different way.

In Theorem B.50 we will see an application of the Multiplicative Möbius Inversion Formula. We conclude this paragraph with another usefull principle.

Theorem A.42 (*Principle of Inclusion and Exclusion*)

Let S be a finite set with N elements. Suppose that the elements in S can satisfy certain properties $P(i)$, $1 \le i \le k$. Let $N(i_1, i_2, \ldots, i_s)$ be number of elements in S that satisfy $P(i_1), P(i_2), \ldots, P(i_s)$, $1 \le i_1 < i_2 < \cdots < i_s \le k$, $1 \le s \le k$, (and possibly also some of the other properties). Let $N(\varnothing)$ denote the number of elements in S that satisfy none of the properties $P(i)$, $1 \le i \le k$. Then

$$N(\varnothing) = N - \sum_{1 \le i \le k} N(i) + \sum_{1 \le i < j \le k} N(i,j) - \cdots + (-1)^k N(1,2, \ldots, k) . \tag{A.53}$$

Proof. An element x in S, that satisfies exactly r properties is counted

$$1 - \begin{bmatrix} r \\ 1 \end{bmatrix} + \begin{bmatrix} r \\ 2 \end{bmatrix} - \cdots + (-1)^r \begin{bmatrix} r \\ r \end{bmatrix} = (1-1)^r = \begin{cases} 1 & \text{if } r = 0, \\ 0 & \text{if } r > 0, \end{cases}$$

times in the right hand side of (A.53), just as in the left hand side. $\Box$

We leave it as an exercise to the reader to prove (A.26) directly from (A.22) with the above principle (Hint: Let $n = \prod_{i=1}^{k} p_i^{e_i}$, $e_i > 0$, $1 \le i \le k$. Take $S = \{1, 2, \ldots, n\}$ and say that an element $j \in S$ has property $P(i)$, $1 \le i \le k$, if j is divisible by p_i.).

Problems

1. Let $p_1^{a_1} p_2^{a_2} \cdots p_k^{a_k}$ be the prime factorization of n. How many different divisors does n have?

2. Compute u and v such that $u \cdot 455 + v \cdot 559 = gcd(455,559)$.

3. a) Proof that 563 is a prime number.
    b) Use Euclid's algorithm to compute 11^{-1} modulo 563.
    c) Solve $11x \equiv 85 \bmod 563$.
    d) Solve $11x \equiv 86 \bmod 563$.

4. Find the solutions of $33x \equiv 255 \bmod 1689$ (note that $1689 = 3 \times 563$ and use the results of problem 3).

5. a) Determine $\phi(100)$.
    b) What are the two least significant digits of 1987^{1987} .

6. Set up the preliminary work necessary to solve systems of equations of the form:

    $$3x \equiv A_1 \bmod 11 ,$$
    $$7x \equiv A_2 \bmod 13 ,$$
    $$4x \equiv A_3 \bmod 15 .$$

    Solve this system of equations for $A_1 = 2$, $A_2 = 9$ and $A_3 = 14$.

7. Determine the Legendre symbol $(7531,3465)$.

8. Use the Chinese Remainder Theorem to solve $x^2 \equiv 56 \bmod 143$.
    How many different solutions are there modulo 143.

9. Proof Corollary A.20 with the Principle of Inclusion and Exclusion and Definition A.13 of $\phi(n)$.

APPENDIX B   THE THEORY OF FINITE FIELDS

§ B.1 Groups, rings, ideals and fields

Let S be a nonempty set. An *operation* $*$ defined on S is a mapping from $S \times S$ into S. The image of the pair (s,t) under $*$ is denoted by $s * t$. The operation $*$ is called *commutative* if

$$\forall_{s,t \in S} \; [s * t = t * s] \,. \tag{B.1}$$

An element e in S that satisfies

$$\forall_{s \in S} \; [s * e = e * s = s] \,, \tag{B.2}$$

will be called a *unit-element* of $(S,*)$. If $(S,*)$ has a unit-element, it will be unique. Indeed, suppose that e and e' both satisfy (B.2). Then by (B.2) $e = e * e' = e'$.

Example B.1 Take $S = \mathbb{Z}$ and $+$ (i.e. addition) as operation. This operation is commutative and $(\mathbb{Z},+)$ has 0 as unit-element.

Example B.2 Let S be the set of 2×2 real matrices and consider matrix multiplication as operation. Then this operation is not commutative, because for instance

$$\begin{bmatrix} 1 & 1 \\ 0 & 1 \end{bmatrix} \begin{bmatrix} 0 & 1 \\ 1 & 0 \end{bmatrix} = \begin{bmatrix} 1 & 1 \\ 1 & 0 \end{bmatrix} \neq \begin{bmatrix} 0 & 1 \\ 1 & 1 \end{bmatrix} = \begin{bmatrix} 0 & 1 \\ 1 & 0 \end{bmatrix} \begin{bmatrix} 1 & 1 \\ 0 & 1 \end{bmatrix} \,.$$

On the other hand S does have a unit-element, namely $\begin{bmatrix} 1 & 0 \\ 0 & 1 \end{bmatrix}$.

Definition B.3 Let G be a non-empty set and $*$ an operation defined on G. Then the pair $(G,*)$ is called a *group*, if

G1:   The operation $*$ is *associative*, i.e.

$$\forall_{g,h,k\in G}\ [g*(h*k)=(g*h)*k].$$

G2: G contains a unit-element, say e.

G3: $\forall_{g\in G}\exists_{h\in G}\ [g*h=h*g=e].$

Property G1 tells us that there is no need to write brackets in strings like $g*h*k$. The element h in Property G3 is unique. Indeed if h and h' both satisfy G3, then $h=h*e=h*(g*h')=(h*g)*h'=e*h'=h'$. The element h satisfying G3 is called the *inverse* of g, and is often denoted by g^{-1}.

A group $(G,*)$ is called commutative, if the operation $*$ is commutative. The reader easily checks that $(\mathbb{Z},+)$ in Example B.1 is a commutative group. Other well-known examples of commutative groups are for instance $(\mathbb{Q},+)$, $(\mathbb{Q}\setminus\{0\},\cdot)$, $(\mathbb{R},+)$, etc..

Example B.2 does not yield a group, because not all matrices have an inverse (e.g. the all-zero matrix).

Let G contain a subset H, such that $(H,*)$ is also a group. In this case $(H,*)$ will be called a *subgroup* of $(G,*)$. Because associativity already holds for the operation $*$, $(H,*)$ will be a subgroup of $(G,*)$ iff

(H1): the product $g*h$ of two elements in H also lies in H,

(H2): the unit-element e lies in H and

(H3): the inverse of an element of H also lies in H.

It is easy to prove that (H2) and (H3) are equivalent to

(H) $\forall_{g,h\in H}\ [g*h^{-1}\in H].$

Let $m\in\mathbb{Z}\setminus\{0\}$. Then $(m\mathbb{Z},+)$, where $m\mathbb{Z}=\{mk\mid k\in\mathbb{Z}\}$, is a commutative subgroup of $(\mathbb{Z},+)$, as one can easily check.

We shall now consider a situation that two operations are defined on a set. The first will be denoted by $g+h$, the second by $g\cdot h$.

Definition B.4 The triple $(R,+,\cdot)$ is called a *ring*, if

R1: $(R,+)$ is a commutative group. Its unit-element will be denoted by 0.

R2:   The operation $\cdot$ is associative.

R3:   *Distributivity* holds, i.e.

$$\forall_{r,s,t\in R}\ [r\cdot(s+t)=r\cdot s+r\cdot t\ \text{ and }\ (r+s)\cdot t=r\cdot t+s\cdot t].$$

From now on we shall simple write gh instead of $g \cdot h$. The (additive) inverse of an element g in the group $(R,+)$ will simply be denoted by $-g$, just as we denote $g + g$ by $2g$, $g + g + g$ by $3g$, etc.. Note that 0 really behaves like a zero-element, because for every $r \in R$ one has that $0r = (r - r)r = r^2 - r^2 = 0$ and similarly that $r0 = 0$.

Suppose that the operation $\cdot$ is commutative on R. Then the ring $(R,+,\cdot)$ will be called commutative. $(\mathbb{Z},+)$, $(\mathbb{Q},+,\ )$, $(\mathbb{R},+,\ )$, but also $(m\mathbb{Z},+,\ )$, $m \neq 0$, are examples of commutative rings.

Let $(R,+,\cdot)$ be a ring and $S \subset R$, such that $(S,+,\cdot)$ is also a ring. Then $(S,+,\cdot)$ is called a *subring*.

Definition B.5 A subring $(S,+,\cdot)$ of a ring $(R,+,\cdot)$ is called an *ideal*, if

I: $\qquad \forall_{r \in R} \ \forall_{s \in S} \ [rs \in S \ \text{ and } \ sr \in S]$.

It is easy to check that an integer multiple of an m-tuple, $m \in \mathbb{Z} \setminus \{0\}$, is also an m-tuple. It follows that $(m\mathbb{Z},+,\ )$ is an ideal in $(\mathbb{Z},+,\ )$.

Now suppose that $(R,\cdot)$ has a unit-element, say e. Then some elements in R may have an inverse in $(R,\cdot)$, i.e. an element b such that $a \cdot b = b \cdot a = e$. This inverse, which is again unique, is called the *multiplicative* inverse of a and will be denoted by a^{-1}. Clearly the element 0 will not have a multiplicative inverse. Indeed suppose that $r0 = e$, for some $r \in R$. Then for each $a \in R$ one has that $a = ae = a(r0) = (ar)0 = 0$, i.e. $R = \{0\}$. It follows that $(R,\cdot)$, $R \neq \{0\}$, can not be a group. However, $(R \setminus \{0\},\cdot)$ may very well have the structure of a group.

Definition B.6 A triple $(F,+,\cdot)$ is called a *field*, if

F1: $(F,+)$ is a commutative group. Its unit-element is denoted by 0.

F2: $(F \setminus \{0\},\cdot)$ is a group. The multiplicative unit-element is denoted by e.

F3: Distributivity holds.

Unlike some rings, a field can not have so-called *zero-divisors*, i.e. elements f and g, both unequal to 0, whose product fg is equal to 0. Indeed suppose that $fg = 0$ and $f \neq 0$. Then $g = e \cdot g = (f^{-1}f)g = f^{-1}(fg) = f^{-1}0 = 0$.

If a subring $(K,+,\cdot)$ of a field $(F,+,\cdot)$ has the structure of a field, we shall call $(K,+,\cdot)$ a *subfield* of $(F,+,\cdot)$.

Examples of fields are the rationals $(\mathbb{Q},+,\ )$, the reals $(\mathbb{R},+,\ )$ and the complex numbers $(\mathbb{C},+,\ )$.

We speak of a *finite* group $(G,*)$, ring $(R,+,\cdot)$ or field $(F,+,\ )$ of *order* n, if G, resp. R or F are finite sets of cardinality n. In this appendix we shall study the structure of finite fields. It will turn out that finite fields of order q only exist when q is a prime power. Moreover these finite fields are essentially unique for a fixed prime power q. This justifies the widely accepted notation $\mathbb{F}_q$ or GF(q) (where GF stands for *Galois Field* after the Frenchman Galois) for a finite field of order q. Examples of finite fields will follow in § B.2.

Analogous to rings we define a commutative field $(F,+, \cdot)$ to be a field, for which $(F\backslash\{0\},\cdot)$ is commutative. The following theorem will not be proved, but is very important [Con77, p. 196].

Theorem B.7 (Wedderburn) Every finite field is commutative.

Definition B.8 Let U be a set. Corresponding to any subset P of $U \times U$, one can define a *relation* ~ on U by

$$\forall_{u,v \in U} [u \sim v \quad \text{iff} \quad (u,v) \in P]. \tag{B.3}$$

An *equivalence* relation is a relation with the additional properties:

E1: $\forall_{u \in U} [u \sim u]$ *(reflexivity)*,

E2: $\forall_{u,v \in U} [u \sim v \Rightarrow v \sim u]$ *(symmetry)*,

E3: $\forall_{u,v,w \in U} [(u \sim v \wedge v \sim w) \Rightarrow u \sim w]$ *(transitivity)*

Let U be the set of straight lines in the (Euclidean) plane. Then "being parallel or equal" defines an equivalence relation. In § A.3 we have seen another example. There $U = \mathbb{Z}$ and the relation was defined by $a \equiv b \mod m$ iff $m \mid (a-b)$.

Let ~ be an equivalence relation defined on a set U. A non-empty subset W of U is called an *equivalence class*, if

i) $\forall_{v,w \in W} [v \sim w]$,

ii) $\forall_{w \in W} \forall_{u \in U\backslash W} [\neg(u \sim w)]$.

It follows from the properties above, that an equivalence class consists of all elements in U, that are in relation ~ with a fixed element in u. Clearly the various equivalence classes of U form a partition of U. The equivalence class containing a particular element w, will be denoted by $<w>$.

Let $(R,+,\cdot)$ be a commutative ring with multiplicative unit-element e. Let $(S,+,\cdot)$ be an ideal in $(R,+,\cdot)$. We define a relation $\equiv$ on R by

$$\forall_{a,b \in R} [a \equiv b \mod S \quad \text{iff} \quad a-b \in S]. \tag{B.4}$$

The reader can easily verify that (B.4) defines an equivalence relation. Let R/S (read: R modulo S) denote the set of equivalence classes. On R/S we define two operations by:

$$<a> + <b> := <a+b>, \qquad a,b \in R, \tag{B.5}$$

$$<a> \cdot <b> := <ab>, \qquad a,b \in R. \tag{B.6}$$

It is easy to verify that these definitions are independent of the particular choice of the elements a and b in the equivalence class $<a>$ resp. $<b>$. We leave it as an exercise to the reader to prove the following theorem.

Theorem B.9 Let $(R,+,\cdot)$ be a commutative ring and let $(S,+,\cdot)$ be an ideal in $(R,+,\cdot)$. With the above definitions $(R/S, +, \cdot)$ is a commutative ring with unit-element.

The ring $(R/S, +, \cdot)$ is called a *residue class ring*. In the next section we will see aplications of Theorem B.9.

Before we conclude this section, there is one more topic that needs to be discussed. Let $(G, \ )$ be a finite group and let a be an element in $G \backslash \{e\}$. Let a^2, a^3, etc. denote $a \cdot a$, $a \cdot a \cdot a$, etc.. Consider the sequence of elements $e, a, a^2, ...$ in G. Since G is finite, there exists a unique integer n such that the elements $e, a, \ldots, a^{n-1}$ are all different, while $a^n = a^j$ for some j, $0 \le j < n$. It follows that $a^{n+1} = a^{j+1}$, etc.. If $j > 0$, it would also follow that $a^{n-1} = a^{j-1}$. However, this contradicts our definition of n. We conclude that $j = 0$, i.e. $a^n = e$. So the elements a^i, $0 \le i \le n$, are all distinct and $a^n = e$.

It is now clear that the elements $e, a, \ldots, a^{n-1}$ form a subgroup H in G. Such a (sub)group H is called a *cyclic* (sub)group of order n. We say that the element a *generates* H and that a has *order* n.

Lemma B.10 Let $(G, \ )$ be a group and a an element in G of order n. Then for all $m \in \mathbb{Z}$

$$a^m = e \quad \text{iff } n \text{ divides } m \ .$$

Proof: Write $m = qn + r$, $0 \le r < n$. Then $a^m = e$ iff $a^r = e$, i.e. iff $r = 0$, i.e. iff $n \mid m$. ☐

Lemma B.11 Let $(G, \ )$ be a group and a an element in G of order n. Then an element a^k, $k > 0$, has order $n/gcd(k, n)$.

Proof: Let m be the order of a^k. Since $k/gcd(k, n)$ is integer, it follows that

$$(a^k)^{n/gcd(k,n)} = (a^n)^{k/gcd(k,n)} = e^{k/gcd(k,n)} = e \ .$$

From Lemma B.10 we conclude that m divides $n/gcd(k, n)$. To prove the converse, we observe that $(a^k)^m = e$. Lemma B.10 implies that n divides km. Hence $n/gcd(k, n)$ divides m. ☐

Analogous to (B.4) one can define for every subgroup $(H, \ )$ of a finite group $(G, \ )$ an equivalence relation $\sim$ by $a \sim b$ iff $ab^{-1} \in H$. The equivalence classes are of the form $\{ha \mid h \in H\}$, as one can easily check. They all have the same cardinality as H. It follows that the number of equivalence classes is $|G|/|H|$. As a consequence $|H|$ divides $|G|$. This proves the following theorem.

Theorem B.12 Let $(G, \ )$ be a finite group of order n. Then every subgroup $(H, \ )$ of $(G, \ )$ has an order dividing n. Also every element $a \ne e$ in G has an order dividing n.

§ B.2 Constructions

The integers modulo m, $m \in \mathbb{N} \setminus \{0\}$, that were introduced in § A.3, can also be described by $(\mathbb{Z} / m\mathbb{Z}, +, \cdot)$ (see Theorem B.9), since $(m\mathbb{Z}, +, \cdot)$ is an ideal in the commutative ring $(\mathbb{Z}, +, \cdot)$. This residue class ring is commutative and has $<1>$ as multiplicative unit-element. The ring $(\mathbb{Z} / m\mathbb{Z}, +, \cdot)$ is often denoted by $(\mathbb{Z}_m, +, \cdot)$.

Theorem B.13 Let m be a positive integer. The ring $(\mathbb{Z}_m, +, \cdot)$ is a finite field with m elements if and only if m is prime.

Proof: Suppose that m is composite, say $m = ab$, $a > 1$ and $b > 1$. Then $<0> = <ab> = <a> \cdot <b>$, while $<a> \neq 0$ and $<b> \neq 0$. So the ring $(\mathbb{Z}_m, +, \cdot)$ has zero-divisors and thus it can not be a field. Now suppose that m is prime. We have to prove that for every equivalence class $<a>$, $<a> \neq <0>$, there exists an equivalence class $<b>$, such that $<a><b> = <1>$. For this it is sufficient to show that for any $a \in \mathbb{Z}$ with $m \nmid a$, there exists an element b, such that $ab \equiv 1 \bmod m$. This however follows from Lemma A.15. $\square$

Later we shall see that for p prime, $(\mathbb{Z}_p, +, \cdot)$ is essentially the only finite field with p elements. We are now going to construct finite fields $(\mathbb{F}_q, +, \cdot)$ for $q = p^m$, p prime.

Let $(F, +, \cdot)$ be a commutative field (not necessarily finite). Let $F[x]$ be the set of *polynomials* over F, i.e. the set of expressions

$$f(x) = f_0 + f_1 x + \cdots + f_n x^n .$$ (B.7)

where $f_i \in F$, $0 \leq i \leq n$, and $n \in \mathbb{N}$. The largest value of i for which $f_i \neq 0$, is called the *degree* of $f(x)$.

Addition and multiplication of polynomials is defined in the natural way.

$$\sum_{i=0}^{n} f_i x^i + \sum_{i=0}^{n} g_i x^i = \sum_{i=0}^{n} (f_i + g_i) x^i .$$ (B.8)

$$\left[\sum_{i=0}^{m} f_i x^i \right] \left[\sum_{j=0}^{n} g_j x^j \right] = \sum_{k=0}^{m+n} \left[\sum_{i+j=k} f_i g_j \right] x^k .$$ (B.9)

It is now straightforward to verify the next theorem.

Theorem B.14 Let $(F, +, \cdot)$ be a commutative field. Then $(F[x], +, \cdot)$ is a commutative ring with unit-element.

Analogous to the concepts defined in App. A for the set of integers, one can define the following notions in $F[x]$: divisibility, *reducibility* (if a polynomial can be written as the product of two polynomials of lower degree), *irreducibility* (which is the analog of primality), *gcd*, *lcm*, unique factorization theorem (the analog of the fundamental theorem in number theory), Euclid's Algorithm, congruence relations, etc.. We leave the details to the reader.

One particular consequence is stated in the following theorem and its corollary.

Theorem B.15 Let $a(x)$ and $b(x)$ be two polynomials in $F[x]$. Then there exist polynomials $u(x)$ and $v(x)$ in $F[x]$, such that

$$u(x)a(x) + v(x)b(x) = gcd(a(x), b(x)) .$$

Corollary B.16 Let $a(x)$ and $f(x)$ be two polynomials in $F[x]$, such that $gcd(a(x), f(x)) = 1$. Then the congruence relation $a(x)u(x) \equiv 1 \bmod f(x)$ has a unique solution modulo $f(x)$.

Another important property of $F[x]$ is given in the following theorem.

Theorem B.17 Any polynomial of degree n, $n > 0$, in $F[x]$ has at most n zeros in F.

Proof: For $n = 1$ the statement is trivial. We proceed by induction on n. Let $u \in F$ be a zero of a polynomial $f(x)$ of degree n over F (if no such u exists, there is nothing to prove). Write $f(x) = (x - u)q(x) + r(x)$, degree$(r(x)) <$ degree$(x - u) = 1$. It follows that $r(x)$ is a constant. Substituting $x = u$ yields $r(x) = 0$. So $f(x) = (x - u)q(x)$. Now $q(x)$ has degree $n - 1$. So by the induction hypothesis, $q(x)$ has at most $n - 1$ zeros in F. Since a field can not have zero-divisors, it follows that $f(x)$ has at most n zeros in F. ☐

Let $s(x) \in F[x]$. It is easy to check that the set $\{a(x)s(x) \mid a(x) \in F[x]\}$ forms an ideal in the ring $(F[x], +,)$. We denote this ideal by $(s(x))$ and say that $s(x)$ *generates* the ideal $(s(x))$.

Conversely let $(S, +,)$ be any ideal in $(F[x], +,)$, $S \neq F[x]$. Let $s(x)$ be a polynomial of lowest degree in S. Take any other polynomial $f(x)$ in S and write $f(x) = q(x)s(x) + r(x)$, degree$(r(x)) <$ degree$(s(x))$. With Properties I and R1, we then have that also $r(x)$ is an element of S. From our assumption on $s(x)$ we conclude that $r(x) = 0$ and thus that $s(x) \mid f(x)$.

It follows from the above discussion that any ideal in the ring $(F[x], +,)$ is generated by a single element! A ring with this property is called a *principal ideal ring*.

From now on we shall restrict ourselves to finite fields. Up to now we have only seen examples of finite fields $I\!F_q$, with q prime.

Let $f(x) \in I\!F_p[x]$. We shall say that f is a *p-ary* polynomial. Let $(f(x))$ be the ideal generated by $f(x)$. By Theorem B.14 we know that $(\mathbf{F}_p[x]/(f(x)), +,)$ is a commutative ring with unit-element $<1>$. It contains p^n elements.

Theorem B.18 Let $(I\!F_p, +,)$ be a finite field with p elements. Let $f(x)$ be a polynomial of degree n over $I\!F_p$. Then the commutative ring $(I\!F_p[x]/(f(x)), +, \cdot)$ is a finite field with p^n elements if and only if $f(x)$ is irreducible in $(I\!F_p[x], +,)$.

Proof: (Compare with the statement and proof of Theorem B.13).

Suppose that $f(x) = a(x)b(x)$, with degree($a(x)$) > 0 and degree($b(x)$) > 0. Then $<a(x)><b(x)> =$
$= <a(x)b(x)> = <f(x)> = <0>$, while $<a(x)> \neq 0$ and $<b(x)> \neq 0$. So $(I\!F_p[x]/(f(x)),+,)$ is a ring
with zero-divisors. Hence it can not be a field.

On the other hand, if $f(x)$ is irreducible, any non-zero polynomial $a(x)$ of degree $< n$ will have a
multiplicative inverse modulo $f(x)$ by Corollary B.16. So $<a(x)><f(x)> = <1>$. Hence
$(I\!F_p[x]/(f(x)),+,)$ is a field.   []

Example B.19 Let $q = 2$. The field $I\!F_2$ consists of the two elements 0 and 1. Let $f(x) = 1 + x + x^3$.
Then $(I\!F_2[x]/(f(x)),+,)$ is a finite field with $2^3 = 8$ elements. These eight elements can be
represented by the eight binary polynomials of degree ≤ 2. The addition and multiplication, as defined
by (B.8) and (B.9), have to be performed modulo $x^3 + x + 1$. For instance

$$(1 + x + x^2)x^2 \equiv x^2 + x^3 + x^4 \equiv (x+1)(x^3 + x + 1) + 1 \equiv 1 \bmod (x^3 + x + 1) .$$

So x^2 is the multiplicative inverse of $1 + x + x^2$ in the field $(I\!F_2[x]/(x^3 + x + 1),+,)$.

Two questions that arise naturally at this moment are:

1)   Does an irreducible, p-ary polynomial $f(x)$ of degree n exist for every prime number p and
      every $n \in I\!N$? If so, then we have proved the existence of finite fields $I\!F_q$ for all prime
      powers q.

2)   Do other finite fields exist?

The first question gets an affirmative answer in the next paragraph. The second question gets a nega-
tive answer in §B.4.

§ B.3 The number of irreducible polynomials over $I\!F_q$

In this section we want to count the number of irreducible polynomials over a finite field $I\!F_q$. Clearly
if $f(x)$ is irreducible, then so is $\alpha f(x)$, for $\alpha \in I\!F_q/\{0\}$. Also the ideals $(f(x))$ and $(\alpha f(x))$ are the
same, when $\alpha \in I\!F_q/\{0\}$. So we shall only count so-called *monic* polynomial of degree n, i.e. polynomials, whose leading coefficient (the coefficient of x^n) is equal to 1.

Definition B.20

$$I_q(n) = \# \ q\text{–ary, irreducible, monic polynomials of degree } n ,$$

$$I(n) = I_2(n) = \# \ \text{binary, irreducible polynomials of degree } n .$$

To develop some intuition for our counting problem, we start with a brute force attack for the special
case that $q = 2$. So we shall try to determine $I(n)$.

There are only two binary polynomials of degree 1, namely

x and $x+1$.

By definition both are irreducible. So $I(1) = 2$.

By taking all possible products of x and $x+1$, one finds three reducible polynomials of degree 2: $x \cdot x = x^2$, $x(x+1) = x^2 + x$ and $(x+1)^2 = x^2 + 1$. Since there are $2^2 = 4$ binary polynomials of degree 2, it follows that there exists only one irreducible polynomial of degree 2, namely

$$x^2 + x + 1 .$$

So $I(2) = 1$.

Each 3-rd degree, reducible, binary polynomial can be written as a product of the lower degree irreducible polynomials x, $x+1$ and $x^2 + x + 1$. In this way one gets $x^i(x+1)^{3-i}$, $0 \le i \le 3$, $(x^2 + x + 1)x$ and $(x^2 + x + 1)(x+1)$. Since there are $2^3 = 8$ binary polynomials of degree 3, we conclude that there are $8 - 4 - 2 = 2$ irreducible, binary polynomials of degree 3. So $I(3) = 2$.

The two binary, irreducible polynomials of degree 3 are:

$$x^3 + x + 1 \quad \text{and} \quad x^3 + x^2 + 1 .$$

At this moment it is important to note that for the counting arguments above, we do not have to know the actual form of the lower degree, irreducible polynomials. We only have to know how many there are of a certain degree.

Indeed to find $I(4)$ we can count the number of reducible, 4-th degree polynomials as follows:

			number
- product of four 1-st degree polynomials			5
- product of one 2-nd degree irreducible polynomial and two			
1-st degree polynomials	1.3	=	3
- product of two 2-nd degree irreducible polynomials			1
- product of one 3-rd degree irreducible polynomial and one			
1-st degree polynomial	2.2	=	4
		total =	13

It follows that there are $2^4 - 13 = 3$ irreducible, binary polynomials of degree 4. So $I(4) = 3$.

With some additional work one can find these three irreducible, 4-th degree polynomials:

$$x^4 + x + 1, \quad x^4 + x^3 + 1 \quad \text{and} \quad x^4 + x^3 + x^2 + x + 1 .$$

Continuing in this way one finds with the necessary perseverance and precision that $I(5) = 6$, $I(6) = 9$, etc.

The above method does not lead to a proof that $I(n) > 0$ for all $n \in I\!N$, let alone to an approximation of the actual value of $I(n)$.

We start all over again.

Let $p_i(x)$, $i = 1,2,...$, be an enumeration of all q-ary, irreducible, monic polynomials, such that for all $i \in \mathbb{N}$, degree($p_i(x)$) $\leq$ degree($p_{i+1}(x)$). So the first $I_q(1)$ polynomials have degree 1, the next $I_q(2)$ polynomials have degree 2, etc..

Any q-ary, monic polynomial $f(x)$ has a unique factorization of the form

$$\prod_{i=1}^{\infty} p_i(x)^{e_i} , \qquad e_i \in \mathbb{N} , \; i \geq 1 ,$$

where only finitely many e_i's are unequal to zero. It follows that $f(x)$ can uniquely be represented by the sequence $(e_1, e_2,...)$. Let a_i be the degree of $p_i(x)$ and n the degree of $f(x)$. Then

$$e_1 a_1 + e_2 a_2 + \cdots = n .$$

So the polynomial $f(x)$ is in a unique correspondence with the term

$$(z^{a_1})^{e_1} (z^{a_2})^{e_2} \cdots$$

in the expression

$$(1 + z^{a_1} + z^{2a_1} + \cdots)(1 + z^{a_2} + z^{2a_2} + \cdots) \cdots ,$$

i.e. in

$$\prod_{i=1}^{\infty} (1 - z^{a_i})^{-1} .$$

Since there are exactly q^n q-ary, monic polynomials of degree n, the above proves that

$$\prod_{i=1}^{\infty} (1 - z^{a_i})^{-1} = 1 + qz + q^2 z^2 + \cdots = (1 - qz)^{-1} ,$$

or equivalently

$$\prod_{i=1}^{\infty} (1 - z^{a_i}) = (1 - qz) . \tag{B.10}$$

From our particular ordering we know that $a_i = k$ for exactly $I_q(k)$ values of i. So (B.10) can be rewritten as:

$$\prod_{k=1}^{\infty} (1 - z^k)^{I_q(k)} = 1 - qz . \tag{B.11}$$

Now take the logarithm of both sides in (B.11) and differentiate. One obtains:

$$q(1 - qz)^{-1} = \sum_{k=1}^{\infty} k I_q(k) z^{k-1} (1 - z^k)^{-1} . \tag{B.12}$$

Multiplying both sides in (B.12) with z yields

$$\sum_{n=1}^{\infty} q^n z^n = qz(1-qz)^{-1} = \sum_{k=1}^{\infty} kI_q(k)z^k(1-z^k)^{-1} =$$

$$= \sum_{k=1}^{\infty} kI_q(k) \sum_{l=1}^{\infty} z^{kl} = \sum_{n=1}^{\infty} \sum_{k \mid n} kI_q(k)z^n .$$

Comparing the coefficients of z^n on both sides gives the relation

$$\sum_{k \mid n} kI_q(k) = q^n . \tag{B.13}$$

Theorem B.21

$$I_q(n) = \frac{1}{n} \sum_{d \mid n} \mu(d)q^{n/d} .$$

Proof: Apply the Möbius Inversion Formula (Theorem A.39) to (B.13).   □

It is now quite easy to determine the asymptotic behaviour of $I_q(n)$ and to prove that its value if always positive.

First of all $I_q(1) = q$, since all monic, polynomials of degree 1 are irreducible by definition. It follows from (B.13) that

$$q + nI_q(n) \le \sum_{k \mid n} kI_q(k) = q^n .$$

Hence

$$I_q(n) \le (q^n - q)/n . \tag{B.14}$$

On the other hand (B.13) and (B.14) imply that

$$q^n = \sum_{k \mid n} kI_q(k) \le nI_q(n) + \sum_{k=0}^{\lfloor n/2 \rfloor} q^k < nI_q(n) + q^{1+n/2} . \tag{B.15}$$

Together with (B.14) this proves the first statement in the following theorem.

Theorem B.22 For all $n \in I\!N$ the number $I_q(n)$ of monic, irreducible, n-th degree polynomials in $I\!F_q[x]$ satisfies

$$\frac{q^n}{n} \left[1 - \frac{1}{q^{n/2-1}} \right] \le I_q(n) \le \frac{q^n}{n} \left[1 - \frac{1}{q^{n-1}} \right] , \tag{B.16}$$

and

$$I_q(n) > 0 . \tag{B.17}$$

Proof: (B.16) is a direct consequence of (B.14) and (B.15). Inequality (B.17) follows from the left hand inequality in (B.16) for $n > 2$. For $n = 1$ and 2 (B.17) follows from Theorem B.21 or directly from $I_q(1) = q > 0$ and $I_q(2) = q^2 - \begin{bmatrix} q+1 \\ 2 \end{bmatrix} = \begin{bmatrix} q \\ 2 \end{bmatrix} > 0$, as one can easily prove directly.   []

Corollary B.23

$$I_q(n) \approx q^n/n \ . \tag{B.18}$$

It follows from this corollary that a randomly selected, monic polynomial of degree n is irreducible with a probability of about $1/n$.

§ B.4 The structure of finite fields

It follows from Theorems B.13, B.18 and B.22, that finite fields $(I\!F_q,+,\cdot)$ exist for all prime powers q. We state this as a theorem.

Theorem B.24 Let p be a prime and $q = p^m$, $m \geq 1$. Then a finite field of order q exists.

Later in this section we shall see that every finite field can be described by the construction of Theorem B.18. But first we shall prove an extremely nice property of finite fields, namely that their multiplicative group is cyclic! By Theorem B.12 we know that every non-zero element in $I\!F_q$ has an order dividing $q - 1$.

Definition B.25 An element a in a finite field of order q is called an *n-th root of unity* if $a^n = e$. An element a is called a *primitive n-th root of unity* if it has order n. If a is a primitive $(q-1)$-st root of unity, then a is called a *primitive element* of $I\!F_q$.

Theorem B.26 Let $(I\!F_q,+,\ )$ be a finite field. Let d be an integer dividing $q - 1$. Then $I\!F_q$ contains exactly $\phi(d)$ elements of order d. In particular $(I\!F_q\backslash\{0\},\ )$ is a cyclic group of order $q - 1$, which contains $\phi(q - 1)$ primitive elements.

Proof: By Theorem B.12 every non-zero element in $I\!F_q$ has a multiplicative order d, which divides $q - 1$. On the other hand, suppose that $I\!F_q$ contains an element of order d, $d \mid (q - 1)$, say a. Then by Theorem B.17 every d-th root of unity in $I\!F_q$ is a power of a.

It follows from Lemma B.11 that $I\!F_q$ contains exactly $\phi(d)$ elements of order d, namely a^i, with $gcd(i,d) = 1$.

Let $a(d)$ be the number of elements of order d in $I\!F_q$. Then the above implies that i) $a(d) = 0$ or $a(d) = \phi(d)$ and ii) $\sum_{d \mid (q-1)} a(d) = q - 1$. On the other hand Theorem A.14 states that $\sum_{d \mid (q-1)} \phi(d) = q - 1$. So $a(d) = \phi(d)$ for all $d \mid (q - 1)$.   []

Corollary B.27 Every element a in $I\!F_q$ satisfies

$$a^{q^n} = a , \qquad n \geq 1 . \tag{B.19}$$

Proof: (B.19) trivially holds for $a = 0$. By Theorem B.12 or Theorem B.26 any $a \neq 0$ has an order dividing $q - 1$. So it satisfies $a^{q-1} = e$. Since $(q-1) \mid (q^n - 1)$, it follows that $a^{q^n-1} = e$, i.e. $a^{q^n} = a$. □

Corollary B.28 Let $I\!\!F_q$ be a finite field. Then

$$x^q - x = \prod_{a \in F_q} (x - a) . \tag{B.20}$$

Proof: Every element a in $I\!\!F_q$ is a zero of $x^q - x$ by Corollary B.27. So the right hand side in (B.20) divides the left hand side. Equality now follows because both sides in (B.20) are monic and of the same degree. □

Example B.29 Consider the finite field $(I\!\!F_2[x]/(f(x)), +, \ )$, with $f(x) = x^4 + x^3 + x^2 + x + 1$. It contains $2^4 = 16$ elements, which can be represented by binary polynomials of degree < 4. The element x, representing the class $\langle x \rangle$, is not a primitive element, since $x^5 \equiv (x+1)f(x) + 1 \equiv 1 \bmod f(x)$. So x has order 5 instead of 15. However, $1 + x$ is a primitive element, as one can verify in Table B.1. Multiplication is easy to perform with Table B.1. For instance

$$(1 + x + x^2 + x^3)(x + x^3) \equiv (1+x)^3 (1+x)^{14} \equiv (1+x)^{17} \equiv$$

$$\equiv (1+x)^2 \equiv 1 + x^2 \bmod f(x) .$$

The element $x + 1$ is a zero of the irreducible polynomial $y^4 + y^3 + 1$, since $1 + (1+x)^3 + (1+x)^4 \equiv 0 \bmod f(x)$. So in $(I\!\!F_2[x]/(g(x)), +, \ )$ with $g(x) = x^4 + x^3 + 1$, the element x is a primitive element. See Table B.2.

Consider the elements $e, 2e, 3e$, etc. in $I\!\!F_q$. Since $I\!\!F_q$ is finite, not all these elements can be different. Also $ie = je$, $i < j$, implies that $(j - i)e = 0$. This justifies the following definition.

Definition B.30 The *characteristic* of a finite field $I\!\!F_q$ with unit-element e, is the smallest positive integer c such that $ce = 0$.

Lemma B.31 The characteristic of a finite field $I\!\!F_q$ is a prime.

Proof: Suppose that the characteristic c can be written as $c'c''$, where $c' > 1$ and $c'' > 1$. Then $0 = ce = (c'e)(c''e)$, while $c'e \neq 0$ and $c''e \neq 0$. So $c'e$ and $c''e$ are zero-divisors. But $I\!\!F_q$ is a field. A contradiction. □

	1	x	x^2	x^3
0	0	0	0	0
$(1+x)^0$	1	0	0	0
$(1+x)^1$	1	1	0	0
$(1+x)^2$	1	0	1	0
$(1+x)^3$	1	1	1	1
$(1+x)^4$	0	1	1	1
$(1+x)^5$	1	0	1	1
$(1+x)^6$	0	0	0	1
$(1+x)^7$	1	1	1	0
$(1+x)^8$	1	0	0	1
$(1+x)^9$	0	0	1	0
$(1+x)^{10}$	0	0	1	1
$(1+x)^{11}$	1	1	0	1
$(1+x)^{12}$	0	1	0	0
$(1+x)^{13}$	0	1	1	0
$(1+x)^{14}$	0	1	0	1

Table B.1 $(I\!F_2[x]/(x^4+x^3+x^2+x+1),+,\ )$, with primitive element $1+x$.

	1	x	x^2	x^3
0	0	0	0	0
1	1	0	0	0
x	0	1	0	0
x^2	0	0	1	0
x^3	0	0	0	1
x^4	1	0	0	1
x^5	1	1	0	1
x^6	1	1	1	1
x^7	1	1	1	0
x^8	0	1	1	1
x^9	1	0	1	0
x^{10}	0	1	0	1
x^{11}	1	0	1	1
x^{12}	1	1	0	0
x^{13}	0	1	1	0
x^{14}	0	0	1	1

Table B.2 $(I\!F_2[x]/(x^4+x^3+1),+,\ )$, with primitive element x.

Definition B.32 Two finite fields $(I\!F_q, +, \cdot)$ and $(I\!F_{q'}, \oplus, \odot)$ are said to be *isomorphic*, if there exists a one-to-one mapping ψ from $I\!F_q$ onto $I\!F_{q'}$ (so $q = q'$), such that for all a and b in $I\!F_q$

$$\psi(a + b) = \psi(a) \oplus \psi(b)$$

$$\psi(ab) = \psi(a) \odot \psi(b) .$$

Lemma B.33 Let $(I\!F_q, +,)$ be a finite field with characteristic p. Then $(I\!F_q, +,)$ contains a subfield which is isomorphic to $(\mathbb{Z}_p, +,)$, i.e. to the integers modulo p.

Proof: The subset $\{ie \mid i = 0, 1, \ldots, p-1\}$ forms a subfield of $(I\!F_q, +,)$, which is isomorphic to $(\mathbb{Z}_p, +,)$ by the (obvious) isomorphism $\psi(ie) = i$, $0 \le i < p$. $\quad\quad\quad\square$

In view of Lemma B.33 we can and shall from now on identify the subfield of order p in $I\!F_q$ with the field $\mathbb{Z}_p$. The subfield $I\!F_p$ is often called the *ground field* of $I\!F_q$. Conversely the field $I\!F_q$ is called an *extension field* of $I\!F_p$.

Theorem B.34 Let $I\!F_q$ be a finite field of characteristic p. Then $q = p^m$, for some integer m, $m \ge 1$.

Proof: Let m be the size of the smallest *basis* of $I\!F_q$ over $I\!F_p$, i.e. m is the smallest integer such that for suitably chosen elements u_i, $1 \le i \le m$, in $I\!F_q$, every element x in $I\!F_q$ can be written as

$$x = \alpha_1 u_1 + \alpha_2 u_2 + \cdots + \alpha_m u_m ,$$

where $\alpha_i \in I\!F_p$, $1 \le i \le m$.
Clearly $q \le p^m$. On the other hand it follows from the minimality of m that different m-tuples $\alpha_1, \alpha_2, \ldots, \alpha_m$ yield different field elements x. So $p^m \le q$. We conclude that $q = p^m$. $\quad\quad\square$

At this moment we know that finite fields $I\!F_q$ can only exist for prime powers q. Theorem B.24 states that $I\!F_q$ indeed does exist for prime powers q. That all finite fields $I\!F_q$ with the same value of q are isomorphic to each other will be proved later.

Theorem B.35 Let a be an element in a finite field $I\!F_q$ of characteristic p. Then in $I\!F_q[x]$

$$(x - a)^p = x^p - a^p . \tag{B.21}$$

Proof: Let $0 < i < p$. Then $gcd(p, i!) = 1$, so

$$\begin{bmatrix} p \\ i \end{bmatrix} \equiv \frac{p(p-1) \cdots (p-i+1)}{i(i-1) \cdots 2 \cdot 1} \equiv 0 \bmod p .$$

And so with the binomial theorem, we have that

$$(x - a)^p = x^p + (-a)^p = x^p - a^p ,$$

where the last equality is obvious for odd p, while for $p = 2$ this equality follows from $+1 = -1$. $\quad\square$

Corollary B.36 Let a_i, $1 \leq i \leq k$, be elements in a finite field $I\!\!F_q$ of characteristic p. Then for every $n \in I\!\!N$

$$\left[\sum_{i=1}^{k} a_i \right]^{p^n} = \sum_{i=1}^{k} a_i^{p^n} .$$

Proof: Use an induction argument on k and on n.   []

The following theorem often gives a powerful criterion to determine, whether an element in a field $I\!\!F_q$ of characteristic p, actually lies in the ground field $I\!\!F_p$.

Theorem B.37 Let $I\!\!F_q$ be a finite field of characteristic p. So $q = p^m$, $m > 0$, and $I\!\!F_q$ contains $I\!\!F_p$ as a subfield. Let a be an element in $I\!\!F_q$. Then a is an element of the subfield $I\!\!F_p$ if and only if a satisfies

$$a^p = a .$$

(B.22)

Proof: The p elements in the subfield $I\!\!F_p$ satisfy (B.22) by Corollary B.27. On the other hand the polynomial $x^p - x$ has at most p zeros in $I\!\!F_q$ by Theorem B.17.   []

Let a be an element in $I\!\!F_q$, a field of characteristic p, but $a \notin I\!\!F_p$. Then $a^p \neq a$ by the previous theorem. Still there is relation between a and a^p.

Theorem B.38 Let a be an element in a finite field $I\!\!F_q$ of characteristic p. Let $f(x)$ be a polynomial over $I\!\!F_p$, such that $f(a) = 0$. Then for all $n \in I\!\!N$

$$f(a^{p^n}) = 0 .$$

(B.23)

Proof: Write $f(x) = \sum_{i=0}^{n} f_i x^i$. Since $f_i \in I\!\!F_p$, $0 \leq i \leq n$, one has by Corollary B.36 and Theorem B.37 that

$$0 = (f(a))^{p^n} = \left[\sum_{i=0}^{n} f_i a^i \right]^{p^n} = \sum_{i=0}^{n} (f_i a^i)^{p^n} =$$

$$= \sum_{i=0}^{n} f_i^{p^n} a^{i \cdot p^n} = \sum_{i=0}^{n} f_i (a^{p^n})^i = f(a^{p^n}) .$$

[]

In $I\!\!R$ and $\mathbb{C}$ a similar thing happens. If $f(x)$ is a polynomial over the reals and $f(a) = 0$, $a \in \mathbb{C}$, then also $f(\overline{a}) = 0$, where $\overline{a}$ is the complex conjugate of a.

The following theorem states that the number of different elements a^{p^i}, $i = 0,1,\ldots$, only depends on the numbers p and the order of a.

Theorem B.39 Let a be an element of order n in a finite field of characteristic p. Let m be the multiplicative order of p modulo n, i.e. $p^m \equiv 1 \bmod n$, with $m \geq 1$ and minimal. Then the m elements

$$a, a^p, a^{p^2}, \ldots, a^{p^{m-1}}$$

are all different and $a^{p^m} = a$.

Proof: By Lemma B.10 (twice) one has that $\alpha^{p^i} = \alpha^{p^j}$ if and only if $p^i \equiv p^j \bmod n$, i.e. iff $p^{i-j} \equiv 1 \bmod n$, i.e. iff $i \equiv j \bmod m$. □

The m elements $a, a^p, \ldots, a^{p^{m-1}}$ in Theorem B.39 are called the *conjugates* of a.

Example B.40 Consider $(\mathbb{F}_2[x]/(f(x)), +, \cdot)$, with $f(x) = x^4 + x^3 + x^2 + x + 1$ (see Example B.29). The element x has order 5. The multiplicative order of 2 modulo 5 is 4. So x, x^2, x^{2^2} and x^{2^3} are all different, while $x^{2^4} = x$. Indeed $x^8 \equiv x^3 \bmod f(x)$, while $x^{16} \equiv x \bmod f(x)$.

Theorem B.41 Let a be an element of order n in a finite field $\mathbb{F}_q$ of characteristic p. Let m be the multiplicative order of p modulo n. Then the polynomial

$$m(x) = \prod_{i=0}^{m-1} (x - a^{p^i}) \tag{B.24}$$

has its coefficients in $\mathbb{F}_p$ and is irreducible in $\mathbb{F}_p[x]$.

Proof: Clearly $m(x)$ is an element in $\mathbb{F}_q[x]$. Write $m(x) = \sum_{i=0}^{m} m_i x^i$. Then by Theorems B.35 and B.39

$$(m(x))^p = \prod_{i=0}^{m-1} (x - a^{p^i})^p = \prod_{i=0}^{m-1} (x^p - a^{p^{i+1}}) =$$

$$= \prod_{i=1}^{m} (x^p - a^{p^i}) = \prod_{i=0}^{m-1} (x^p - a^{p^i}) = m(x^p).$$

Hence

$$\sum_{i=0}^{m} m_i x^{pi} = m(x^p) = (m(x))^p = \left[\sum_{i=0}^{m} m_i x^i \right]^p = \sum_{i=0}^{m} m_i^p x^{pi}.$$

Comparing the coefficients of x^{pi} on both hands yields $m_i = m_i^p$. With Theorem B.37 we then have that $m_i \in \mathbb{F}_p$, $0 \leq i \leq m$. So $m(x)$ is a polynomial in $\mathbb{F}_p[x]$. From Theorems B.38 and B.39 it follows that no polynomial in $\mathbb{F}_p[x]$ of degree less than m can have a as a zero. So $m(x)$ is irreducible in $\mathbb{F}_p[x]$. □

Corollary B.42 Let a be an element of order n in a finite field of characteristic p. Let $m(x)$ be defined by (B.24) and let $f(x)$ be any p-ary polynomial that has a as zero. Then $f(x)$ is a multiple of $m(x)$.

Proof: Combine Theorems B.38, B.39 and B.41.   ▯

So $m(x)$ is the monic polynomial of lowest degree over $\mathbb{F}_p$, having a as a zero. The polynomial $m(x)$ is called the *minimal polynomial* of a. It has a and all the conjugates of a as zeros. The degree of the minimal polynomial $m(x)$ of an element a, is often simply called the *degree* of a.

If $m(x)$ is the minimal polynomial of a primitive element, then $m(x)$ is called a *primitive polynomial*. Let $m(x)$ be the minimal polynomial of an element a of degree m. It follows from Corollary B.42 that the p^m expressions $\sum_{i=0}^{m-1} f_i a^i, f_i \in \mathbb{F}_p, 0 \le i \le m$, take on p^m different values. For these expressions addition and multiplication can be performed, just as in (B.8) and (B.9), where $m(a) = 0$ has to be used to reduce the degree of the outcome to a value less than m. It is quite easy to check that one obtains a field, that is isomorphic to $\mathbb{F}_p[x]/(m(x))$. If $m(x)$ is primitive, one has that $1, x, \ldots, x^{p^m-2}$ are all different modulo $m(x)$, just as the elements $1, a, \ldots, a^{p^m-2}$ are all different. See for instance Example B.29, where the primitive element $a = 1+x$ has minimal polynomial $m(y) = 1+y^3+y^4$. Table B.2 shows the field $\mathbb{F}_2[y]/(m(y))$.

Lemma B.43 Let $m(x)$ be an irreducible polynomial of degree m over a field with p elements and let n be a multiple of m. Then $m(x)$ divides $x^{p^n} - x$.

Proof: Consider the residue class ring $\mathbb{F}_p[x]/(m(x))$. This ring is a field with $q = p^m$ elements by Theorem B.18. The field element $<x>$ in $\mathbb{F}_q$ is a zero of $m(x)$, since $m(<x>) = <m(x)> = <0>$. It follows from Corollary B.27 that $<x>$ is a zero of $x^{p^n} - x$, $i \ge 1$. By Corollary B.42 $m(x)$ divides $x^{p^n} - x$.   ▯

Also the converse of Lemma B.43 is true.

Theorem B.44 In $\mathbb{F}_p[x]$ the polynomial $x^{p^n} - x$ is the product of all irreducible, monic, p-ary polynomials of a degree dividing n.

Proof: Let $m \mid n$. There are $I_p(m)$ irreducible polynomials in $\mathbb{F}_p[x]$, all of which divide $x^{p^n} - x$ by Lemma B.43. The sum of their degree is $m I_p(m)$. Since $\sum_{m \mid n} m I_p(m) = p^n = \text{degree}(x^{p^n} - x)$ by (B.13), it follows that the irreducible, monic, p-ary polynomials of degree m, $m \mid n$, form the complete factorization of $x^{p^n} - x$.   ▯

Example B.45 $p = 2, n = 4, I_2(1) = 2, I_2(2) = 1$ and $I_2(4) = 3$ (see § B.2).

$$x^{16}-x = x(x+1)(x^2+x+1)(x^4+x+1)(x^4+x^3+1)(x^4+x^3+x^2+x+1) \ .$$

Corollary B.46 Let $f(x)$ be an irreducible polynomial in $I\!\!F_p[x]$ of degree m. Let $m \mid n$. Then a finite field with p^n elements contains m roots of $f(x)$.

Proof: By Theorem B.44 $f(x)$ divides $x^q - x$, $q = p^n$. On the other hand $x^q - x = \prod\limits_{a \in F_q} (x-a)$, by (B.20). $\Box$

Theorem B.47 Let p be a prime and m in $I\!\!N$. Then the finite field $I\!\!F_{p^m}$ is unique, up to isomorphism.

Proof: Write $q = p^m$ and let $I\!\!F_q$ be any finite field of order q. Let $f(x)$ be any irreducible, p-ary polynomial of degree m. We shall show that $I\!\!F_q$ is isomorphic to $I\!\!F_p[x]/(f(x))$.

By Corollary B.46 $I\!\!F_q$ contains m zeros of $f(x)$. Let a be one of these m zeros. Since $f(x)$ is irreducible in $I\!\!F_p[x]$, we know that $1, a, \ldots, a^{m-1}$ are independent over $I\!\!F_p$. So any element in $I\!\!F_q$ can be written as $\sum\limits_{i=0}^{m-1} f_i a^i, f_i \in I\!\!F_p, 0 \le i \le m-1$.

The isomorphism between $I\!\!F_q$ and $I\!\!F_p[x]/(f(x))$ is now obvious. $\Box$

Corollary B.48 $I\!\!F_{p^m}$ is (isomorphic to) a subfield of $I\!\!F_{p^n}$ if and only if m divides n.

Proof: The following assertions are all equivalent:

i) $m \mid n$,

ii) $(p^m - 1) \mid (p^n - 1)$,

iii) $x^{p^m} - x$ divides $x^{p^n} - x$,

iv) $\prod\limits_{a \in I\!\!F_{p^m}} (x-a)$ divides $\prod\limits_{a \in I\!\!F_{p^n}} (x-a)$,

v) $I\!\!F_{p^m}$ is a subfield of $I\!\!F_{p^n}$. $\Box$

Example B.49 It follows from Corollary B.48 that $I\!\!F_{2^4}$ contains $I\!\!F_{2^2}$ as a subfield, but does not contain $I\!\!F_{2^3}$ as a subfield. From Table B.2 one can easily verify that the elements $0, 1, x^5$ and x^{10} form a subfield of order 2^2 in $I\!\!F_2[x]/(x^4+x^3+1)$.

Consider a finite field $I\!\!F_q$ of characteristic p. So $q = p^m$ for some $m > 0$. By Theorem B.12 every element in $I\!\!F_q$ has an order dividing $q-1$. Let a be a primitive n-th root of unity in $I\!\!F_q$. So $n \mid (q-1)$. Let $d \mid n$ and $b = a^{n/d}$. Then b is a primitive d-root of unity. Clearly the d elements $1, b, \ldots, b^{d-1}$ satisfy $x^d = 1$. By Theorem B.17 no other element in $I\!\!F_q$ satisfies $x^d = 1$. In $I\!\!F_q$ we define for any $d \mid (q-1)$ the *cyclotomic polynomial* $Q^{(d)}$ by

$$Q^{(d)}(x) = \prod_{\substack{\alpha \in \mathbf{F}_q \\ \alpha \text{ has order } d}} (x - \alpha) . \tag{B.25}$$

By Theorem B.26 $Q^{(d)}(x)$ has degree $\phi(d)$.

Since a is a primitive n-th root of unity, it follows that

$$x^n - 1 = \prod_{i=0}^{n-1} (x - a^i) = \prod_{\substack{\alpha \\ \alpha \text{ has an order dividing } n}} (x - \alpha) =$$

$$= \prod_{d \mid n} \prod_{\substack{\alpha \\ \alpha \text{ has order } d}} (x - \alpha) = \prod_{d \mid n} Q^{(d)}(x) . \tag{B.26}$$

Theorem B.50

$$Q^{(d)}(x) = \prod_{d \mid n} (x^d - 1)^{\mu(n/d)} . \tag{B.27}$$

Proof: Apply the Multiplicative Möbius Inversion Formula (Corollary A.40) to (B.26).

Example B.51 By (B.27)

$$Q^{(36)}(x) = \frac{(x^{36} - 1)(x^6 - 1)}{(x^{18} - 1)(x^{12} - 1)} = \frac{x^{18} + 1}{x^6 + 1} = x^{12} - x^6 + 1 .$$

All the irreducible factors of $Q^{(d)}(x)$ have the same degree, because all the zeros of $Q^{(d)}(x)$ have the same order d. Indeed, by Theorem B.39 each irreducible factor of $Q^{(d)}$ has as degree the multiplicative order of p mod d. In particular we have the following theorem.

Theorem B.52 The number of primitive, p-ary, monic polynomials of degree m is

$$\phi(p^m - 1)/m .$$

Proof: A primitive, p-ary polynomial of degree m divides $Q^{(p^m - 1)}(x)$ and this cyclotomic polynomial has no other factors. The degree of $Q^{(p^m - 1)}(x)$ is $\phi(p^m - 1)$.

Example B.53 $p = 2$.

$$x^{16} - x = x(x^{15} - 1) = xQ^{(1)}(x)Q^{(3)}(x)Q^{(5)}(x)Q^{15}(x) ,$$

where

$$Q^{(1)}(x) = x + 1 ,$$

$$Q^{(3)}(x) = x^2 + x + 1 ,$$

$$Q^{(5)}(x) = x^4 + x^3 + x^2 + x + 1 ,$$

$$Q^{(15)}(x) = (x^4 + x + 1)(x^4 + x^3 + 1) .$$

Indeed there are $\phi(15)/4 = 2$ primitive polynomials of degree 4. See also Example B.29.

Most of the time in this chapter we have viewed $I\!\!F_q$, $q = p^m$ and p prime, as an extension field of $I\!\!F_p$. But all the concepts, defined in this chapter, can also be generalized to $I\!\!F_q[x]$. So one may want to count the number of irreducible polynomials of degree n in $I\!\!F_q[x]$ or discuss primitive polynomials over $I\!\!F_q$, etc. We leave it to the reader to verify that all the theorems in this appendix can indeed be generalized from $I\!\!F_p$ and $I\!\!F_{p^m}$ to $I\!\!F_q$ resp. $I\!\!F_{q^m}$, simply by replacing p by q and q by q^m.

Example B.55 The field $I\!\!F_{16}$ can be viewed as the residue class ring $I\!\!F_4[x]/(x^2+x+\alpha)$, where α is an element in $I\!\!F_4$, satisfying $\alpha^2 + \alpha + 1 = 0$.

Problems

1. Proof that $(\{x \in R \mid x^2 \in \mathcal{Q} , x = 0,\} \cdot)$ is a group.

2. Proof that the elements of a reduced residue class system modulo m form a multiplicative group.

3. Proof that there are essentially two different groups of order 4 (hint: each element has an order dividing 4).

4. Find an element of order 12 in the group $(Z_{13}\backslash\{0\} , \cdot)$. Which powers of this element have order 12. Answer the same question for elements of order 6, 4, 3, 2 and 1.

5. Use Euclid's Algorithm to find the multiplicative inverse of $1 + x + x^2$ mod $1 + x^2 + x^4$.

6. How many binary, irreducible polynomials are there of degree 7 and 8?

7. Make a log table of $GF(2)[x] / (1+x^2+x^5)$ (hint: x is a primitive element).

8. Which subfields are contained in GF(625). Let α be a primitive elelement in GF(625). Which powers of α constitute the various subfields of GF(625).

9. Proof that over GF(2)

$$(x + y)^{2^k + 1} = x^{2^k + 1} + x^{2^k} y + xy^{2^k} + y^{2^k + 1} \, .$$

10. How many primitive polynomials are there of degree 10?

11. Determine $Q^{(21)}(x)$. What is the degree of the factors of $Q^{(21)}(x)$.

12. What is the degree of a binary minimal polynomial of a primitive 17-th root of unity? How many such polynomials do exist? Prove that each is its own reciprocal. Determine them explicitly.

13. The *trace* mapping Tr is defined on $GF(p^m)$, p prime, by

$$\mathrm{Tr}(x) = x + x^p + x^{p^2} + \cdots + x^{p^{m-1}} \, .$$

a) Prove that $\mathrm{Tr}(x) \in GF(p)$, for every $x \in GF(p^m)$. So Tr is a mapping from $GF(p^m)$ to $GF(p)$.

b) Prove that Tr is a linear mapping.

c) Prove that Tr takes on *every* value in $GF(p)$ equally often (hint: use Theorem B.17).

d) Replace p by q in this problem, where q is a prime power, and verify the same statements.

REFERENCES

[Adl79] Adleman, L.M., A subexponential algorithm for the discrete logarithm problem with applications to cryptography, in Proc. IEEE 20-th Annual Symp. on Found. of Comp. Science, pp. 55-60, 1979.

[Adl83] Adleman, L.M., On breaking the iterated Merkle-Hellman public key cryptosystem, in Proc. 15-th Annual ACM Symp. Theory of Computing, pp. 402-412, 1983.

[Adl83] Adleman, L.M., C. Pomerance and R. Rumely, On distinguishing prime numbers from composite numbers, Annals of Math., 17, pp. 173-206, 1983.

[Aig79] Aigner, M., Combinatorial Theory, Springer Verlag, Berlin, etc., 1979.

[Bek82] Beker, H. and F. Piper, Cipher systems, the protection of communications, Northwood Books, London, 1982.

[Ber68] Berlekamp, E.R., Algebraic coding theory, McGraw-Hill Book Company, New York, etc., 1968.

[Ber78] Berlekamp, E.R., R.J. McEliece and H.C.A. van Tilborg, On the inherent intractability of certain coding problems, IEEE Trans. Inf. Theory, IT-24, pp. 384-386, May 1978.

[Bla84] Blake, I.F., R. Fuji-Hara, R.C. Mullin and S.A. Vanstone, Computing logarithms in finite fields of characteristic two, SIAM J. Algebraic and Disc. Meth., 15, pp. 276-285, June 1984.

[Blu86]   Blum, M., How to prove a theorem so no one else can claim it, presented at the International Congress of Mathematicians at Berkeley Calif., 1986.

[Bri85]   Brickell, E.F., Breaking iterated knapsacks, in Advances in Cryptography: Proc. of Crypto '84, G.R. Blakley and D. Chaum, Eds., Lecture Notes in Computer Science 196, Springer Verlag, Berlin etc., pp. 342-358, 1958.

[Cha85]   Chaum, D., Security without identification: transaction systems to make big brother obsolete, Communications of the ACM, $\underline{28}$, pp. 1030-1044, 1985.

[ChE87]   Chaum, D. and J.-H. Evertse, A secure and privacy protecting protocol for transmitting personal information between organizations, Advances in Cryptology - Crypto '86, A.M. Odlyzko, Ed., Lecture Notes in Computer Science 263, Springer, Berlin, pp. 118-167, 1987.

[ChG85]   Chor, B. and O. Goldreich, RSA/Rabin least signigicant bits are $\dfrac{1}{2} + \dfrac{1}{\text{poly}(\log N)}$ secure, in Advances in Cryptography: Proc. of Crypto '84, G.R. Blakley and D. Chaum, Eds., Lecture Notes in Computer Science 196, Springer Verlag, Berlin etc., pp. 303-313, 1985.

[ChR85]   Chor, B. and R.L. Rivest, A knapsack type public key cryptosystem based on arithmetic in finite fields, in Advances in Cryptography: Proc. of Crypto '84, G.R. Blakley and D. Chaum, Eds., Lecture Notes in Computer Science 196, Springer Verlag, Berlin etc., pp. 54-65, 1985.

[Coh82]   Cohen, H. and H.W. Lenstra Jr., Primality testing and Jacobi sums, Report 82-18, Math. Inst., Univ. of Amsterdam, Oct. 1982.

[Con77]   Cohn, P.M. Algebra Vol.2, John Wiley & Sons, London, etc., 1977.

[Cop84]   Coppersmith, D., Fast evaluation of logarithms in fields of characteristic two, IEEE Trans. Inf. Theory, IT-$\underline{30}$, pp. 587-594, July 1984.

[Cov67]   Coveyou, R.R. and R.D. Mcpherson, Fourier analysis of uniform random number generators, J. Assoc. Comput. Mach., $\underline{14}$, pp. 100-119, 1967.

[Den82]   Denning, D.E.R., Cryptography and data security, Addison-Wesley Publ. Comp., Reading Ma, etc., 1982.

[Des86]   Desmedt, Y., An exhaustive key search machine breaking one million DES keys, presented at Eurocrypt '86, Linköping, 1986.

[Dif76]   Diffie, W. and M.E. Hellman, New directions in cryptography, IEEE Trans. Inf. Theory, IT-$\underline{22}$, pp. 644-654, Nov. 1976.

[DiH76]   Diffie, W. and M.E. Hellman, A critique of the proposed Data Encryption Standard. Comm. ACM, $\underline{19}$, pp. 164-165, 1976.

[DoY81]   Dolev, D. and A.C. Yao, On the security of public key protocols, Proc. 22-nd Ann. FOCS Symp., pp. 350-357, 1981.

[Eli87]   Elias, P., Interval and recency-rank source coding: two on-line adaptive variable-length schemes, IEEE Trans. Inf. Theory, IT-$\underline{33}$, pp. 3-10, Jan. 1987.

[Fri73]   Friedman, W.F., Cryptology, in Encyclopedia Brittanica, p. 848, 1973.

[Gar79]   Garey, M.R. and D.S. Johnson, Computers and intractability: A guide to the theory of NP-completeness, W.H. Freeman and Co., San Francisco, 1979.

[GMR85]   Goldwasser S., S. Micali and C. Rackoff, The knowledge complexity of interactive proof-systems, Proc. of 17th Annual ACM Symp. on Theory of Comp. ACM, Providence, pp. 291-304, 1985.

[GMi84]   Goldwasser, S. and S. Micali, Probabilistic encryption, Jour. of Computer and System Science, $\underline{28}$, pp. 270-299, 1984.

[GMT82]   Goldwasser, S., S. Micali and P. Tong, Why and how to establish a private code over a public network, Proc. FOCS, pp. 112-117, 1982.

[Gol67]   Golomb, S.W., Shift register sequences, Holden-Day, San Francisco, 1967.

[Har45]   Hardy, G.H. and E.M. Wright, An introduction to the theory of numbers, Clarendon Press, Oxford, 1945.

[Hel77]   Hellman, M.E.,An extension of the Shannon theory approach to cryptography, IEEE Trans. Inf. Theory, IT-$\underline{23}$, pp. 289-294, May 1977

[Hel79]   Hellman, M.E., DES will be totally insecure within ten years, IEEE Spectrum. $\underline{16}$, pp .32-39, 1979.

[Hel83]   Hellman, M.E. and J.M. Reyneri, Fast computation of discrete logarithms over GF(q), in Advances in cryptography: Proc. of Crypto '82, D. Chaum, R. Rivest and A. Sherman, Eds., Plenum Publ. Comp., New York, pp. 3-13, 1983.

[Huf52]   Huffmann, D.A., A method for the construction of minimum-redundancy codes, Proc. IRE, $\underline{14}$, pp.1098-1101, 1952.

[Jen83]   Jennings, S.M., Multiplexed sequences: some properties of the minimum polynomial, in: Cryptography Proc. Burg Feuerstein 1982 (ed. by Th. Beth), Lecture Notes in Computer Science 149, Springer Verlag, Berlin etc., pp. 189-206, 1983.

[Kah67]   Kahn, D., The codebreakers. the story of secret writing, Macmillan Company, New York, 1967.

[Khi57]   Khinchin. A.I., Mathematical foundations of information theory, Dover Publications, New York, 1957.

[Knu69]   Knuth, D.E., The art of computer progamming, Vol.2, Semi-Numerical Algorithms, Addison-Wesley, Reading, MA., 1969.

[Knu73]   Knuth, D.E., The art of computer progamming, Vol.3, Sorting and searching, Addison-Wesley, Reading, MA., 1973.

[Knu81]   Knuth, D.E., The art of computer programming, Vol.2, Semi-Numerical Algorithms, Second Edition, Addison-Wesley, Reading, MA., 1981.

[Kon81]   Konheim, A.G., Cryptography, a primer, John Wiley & Sons, New York, etc., 1981.

[Kra49]   Kraft, L.G. A device for quantizing, grouping and coding amplitude modulated pulses, MS Thesis, Dept. of EE, MIT, Cambridge, Mass., 1949.

[Lag83]   Lagarias, J.C. and A.M. Odlyzko, Solving low-density subset problems, Proc. 24th Annual IEEE Symp. on Found. of Comp. Science, pp.1-10, 1983

[Lag84]   Lagarias, J.C., Knapsack-type public key cryptosystems and Diophantine approximation, in Advances in Cryptography: Proc. of Crypto '83, D. Chaum ed., Plenum Publ. Comp., New York, pp.3-23, 1984.

[Leh76]   Lehmer, D.H., Strong Carmichael numbers, J. Austral. Math. Soc., Ser. A 21, pp. 508-510, 1976.

[Len82]   Lenstra, A.K., H.W. Lenstra, Jr. and L. Lovász, Factoring polynomials with rational coefficients, Mat. Annalen, 261, pp. 515-534, 1982.

[Len83]   Lenstra, H.W. Jr., Fast prime number tests, Nieuw Archief voor Wiskunde (4), 1, pp. 133-144, 1983.

[Len86]   Lenstra, H.W. Jr., Factoring integers with elliptic curves, Report 86-16, Dept. of Mathematics, University of Amsterdam, Amsterdam, the Netherlands.

[Liu68]   Liu, C.L., Introduction to combinatorial mathematics, McGraw-Hill, New-York, 1968.

[Lün87]   Lüneburg, On the rational normal form of endomorphisms; a primer to constructive algebra, BI Wissenschaftsverlag, Mannheim etc., 1987.

[Mac77]   MacWilliams, F.J. and N.J.A. Sloane, The theory of error-correcting codes, North-Holland Publ. Comp., Amsterdam, etc., 1977.

[Mas69]   Massey, J.L., Shift-register synthesis and BCH decoding, IEEE Trans. Inf. Theory, IT-15, pp.122-127, Jan. 1969.

[McE78]   McEliece, R.J., A public-key cryptosystem based on algebraic coding theory, JPL DSN Progress Report 42-44, pp.114-116, Jan.-Febr. 1978.

[McE81] McEliece, R.J. and D.V. Sarwate, On sharing secrets and Reed-Solomon codes, Comm. ACM, Vol. 24, pp. 583-584, Sept. 1981.

[McM56] McMillan, B., Two inequalities implied by unique decipherability, IEEE Trans. Inf. Theory, IT-56, pp.115-116, Dec. 1956.

[Mer78] Merkle, R.C. and M.E. Hellman, Hiding information and signatures in trapdoor knapsacks, IEEE Trans. Inf. Theory, IT-24, pp. 525-530, Sept. 1978.

[Mey82] Meyer, C.H. and S.M. Mathyas, Cryptography: a new dimension in computer data security, John Wiley & Sons, New York, etc., 1982.

[MoB75] Morrison, M.A. and J. Brillhart, A method of factoring and the factorization of F_7, Math. Comp. 29, pp.183- 205, 1975.

[Mor77] Morris, R., N.J.A. Sloane and A.D. Wyner, Assessment of the National Bureau of Standards proposed Federal Data Encryption Standard, Cryptologia 1, pp. 281-306, 1977.

[Odl85] Odlyzko, A.M., Discrete logarithms in finite fields and their cryptographic significance, Advances in Cryptology: Proc. Eurocrypt '84, T. Beth, N. Cot and I. Ingemarsson, Eds., Lecture Notes in Computer Science 209, Springer, Berlin etc., pp. 224-314, 1985.

[Pat75] Patterson, N.J., The algebraic decoding of Goppa codes, IEEE Trans. Inf. Theory, IT-21, pp. 203-207, Mar. 1975.

[Per86] Peralta, R., A simple and fast probablistic algorithm for computing square roots modulo a prime number, presented at Eurocrypt '86, Linköping, 1986.

[Poh78] Pohlig, S.C. and M.E. Hellman, An improved algorithm for computing logarithms over GF(p) and its cryptographic significance, IEEE Trans. Inf. Theory, IT-24, pp. 106-110, Jan. 1978

[Pol75] Pollard, J.H., A Monte Carlo method for factoring, BIT 15, pp. 331-334, 1975.

[Rab79] Rabin, M.O., Digitalized signatures and public-key functions as intractable as factorization, MIT/LCS/TR-212, MIT Lab. for Comp. Science, Cambridge, Mass., Jan. 1979.

[Rab80] Rabin, M.O., Probabilistic algorithms in finite fields, SIAM J. Comput. 80, pp. 273-280, 1980.

[Riv78] Rivest, R.L., A. Shamir and L. Adleman, A method for obtaining digital signatures and public key cryptosystems, Comm. ACM, Vol. 21, pp. 120-126, Febr. 1978.

[Riv83] Rivest, R.L. and A.T. Sherman, Randomized encryption techniques, in Advances in Cryptology, Proc. of Crypto '82, D. Chaum, R. Rivest and A.T. Sherman, Eds., Plenum Publ. Comp., New York, pp. 145-163, 1983.

[Rue86]   Rueppel, R.A., Analysis and design of streamciphers, Springer-Verlag, Berlin etc., 1986.

[Sha79]   Shamir, A., How to share a secret, Comm. ACM., Vol. $\underline{22}$, pp. 612-613, Nov. 1979.

[Sha82]   Shamir, A., A polynomial time algorithm for breaking the basic Mekle-Hellman cryptosys-
           tem, in Proc. 23-rd IEEE Symp. Found. Computer Sci., pp. 145-152, 1982.

[Shn49]   Shannon, C.E., Communication Theory and Secrecy Systems, B.S.T.J., $\underline{28}$, pp. 656-715,
           Oct. 1949.

[Shp83]   Shapiro, H.N., Introduction to the theory of numbers, John Wiley & Sons, New York, etc.,
           1983.

[Sol77]   Solovay, R. and V. Strassen, A fast Monte-Carlo test for primality, SIAM J. Comput. $\underline{6}$, pp.
           84-85, March 1977.

[Sug76]   Sugiyama, Y., M. Kasahara, S. Hirasawa and T. Namekawa, An erasures-and-errors decod-
           ing algorithm for Goppa codes, IEEE Trans. Inf. Theory, IT-$\underline{22}$, pp. 238-241, Mar. 1976.

[Wil86]   Willems, F.M.J., Universal data compression and repetition times, in Proceedings of the
           Seventh Symposium on Information Theory in the Benelux, D.E. Boekee ed., Delft
           University Press, Delft, pp. 73-80, 1986.

[Wlm79]   Williams, H.C. and B. Schmid, Some remarks concerning the M.I.T. public-key cryptosys-
           tem, BIT $\underline{19}$, pp. 525-538, 1979.

[Ziv77]   Ziv, J. and A. Lempel, A universal algorithm for sequential data compression, IEEE Trans.
           Inf. Theory, IT-$\underline{22}$, pp. 337-343, May 1977.

NOTATIONS

$a \mid b$	a divides b	119
$a \nmid b$	a does not divide b	119
(a,b)	greatest common divisor of a and b	120, 142
$[a,b]$	least common multiple of a and b	120, 142
$\mid A \mid$	cardinality of set A	1
(a/m)	Legendre symbol, Jacobi symbol	82, 129
$\chi(a)$	Legendre symbol	129
$\mu(n)$	Möbius function	132
$a \underline{\bmod} n$	nonnegative remainder of a divided by b	7
$O(f(n))$	Landau symbol	67
A^n	set of n-grams of the alphabet A	2
A^*	$\cup_{n \geq 0} A^n$	2
$m \mathbb{Z}$	the set of multiples of m in $\mathbb{Z}$	139
$\mathbb{Z}_m$	the integers modulo m	142
$\mathbb{F}_q$	finite field with q elements	139
$F[x]$	ring of polynomials in x over F	142
$Q^{(d)}(x)$	cyclotomic polynomial	156
$h(p), h(\underline{p})$	entropy	40
$H(x)$	entropy	40
$H(X \mid Y)$	equivocation, conditional entropy	43
$I(X;Y)$	mutual information	44

INDEX

active cryptanalysis — 5
Adleman Algorithm — 73
Algorithm
- Adleman — 73
- Berlekamp-Massey — 32
- Euclid's — 121, 143
- Huffman — 51
- Pohlig-Hellman — 69
- prime number generator — 81
- Shortest Vector — 107
alphabet — 1
associativity — 138
attack
- ciphertext only — 6
- chosen plaintext — 6
- known plaintext — 6
authentication — 1
autocorrelation — 20
- in-phase — 20
- out-of-phase — 20
average redundancy per letter — 42

basis — 105, 151
Berlekamp-Massey algorithm — 32
binary symmetric channel — 44
bit — 39
bit security — 81
block — 20
block cipher — 16

branch point — 33
BSC — 44

cardinality — 1
Caesar cipher — 7
characteristic — 149
characteristic polynomial — 24
Chinese Remainder Theorem — 127
chosen plaintext attack — 6
cipher block chaining — 56
cipher system
- block — 16
- stream — 16
ciphertext — 5
ciphertext only attack — 6
code
- Goppa — 93
- Huffman — 52
- instantaneous — 48
- prefix — 48
- Reed-Solomon — 112
- uniquely decodable — 47
Cohen and Lenstra primality test — 84, 88
coincidence — 11
column transposition — 15
commutativity — 137
complete residue system — 124
conditional entropy — 42
confusion — 16

congruent	124
conjugates	153
conventional cryptosystem	1
credential mechanism	116
cryptanalysis	1
- active	5
- passive	5
cryptography	1
cryptographic transformation	5
cryptology	1
cryptosystem	1, 5
- Caesar	7
- column transposition	15
- conventional	1
- DES	55
- discrete logarithm problem	68
- Enigma	16
- feedback shift register	22
- Hagelin	16
- knapsack	98
- McEliece	94
- multiplexing	34
- one-time pad	14
- Playfair	15
- polyalphabetic substitution	11
- product	16
- public key	63
- Rabin's variant to R.S.A.	89
- randomized	115
- rotor machines	16
- R.S.A.	77
- shift register	22
- simple substitution	8, 42
- transposition	15
- Vernam	14, 46
- Vigenère	9
cyclic group	141
cyclotomic polynomial	155
data compression	47
Data Encryption Standard	55
deciphering	5
decryption	5
degree	142, 154
density	105
DES	55
discrete logarithm problem	67
diffusion	16
distributivity	138
divide	119
eavesdropping	5
enciphering	5
encryption	5
Enigma	16
entropy	40
entropy (conditional)	43
equilibrium distribution	4
equivalence class	140
equivalence relation	140
equivocation	43
Euclid	119
Euclidean norm	106
Euclid's Algorithm	121, 143
Euler's Theorem	77, 125
Euler's totient function	26, 77, 125
exhaustive keysearch	7
extension field	151
feedback coefficients	23
feedback function	22
feedback shift register	22
Fermat's Theorem	85, 126
Fibonacci numbers	123
field	139
finite	
- field	139
- group	139
- ring	139
Fundamental Theorem of Number Theory	121
Galois field	139
gap	20
gcd	120, 142
generate	141, 143
generating function	25
GF	139
Golomb's Randomness Postulates	20
Goppa code	93
Gram-Schmidt process	106
greatest common divisor	120, 142
ground field	151
group	138
Hagelin's cipher machine	16
hot line	14
Huffman Algorithm	51
Huffman code	52
ideal	139
incidence of coincidences	11
inclusion and exclusion (principle of)	134
information	39
information (mutual)	44
in-phase autocorrelation	20

instantaneous code	48
integer lattice	105
inverse	138
irreducible polynomial	26, 142
isomorphic	151
Jacobi symbol	82, 131
Kasiski, F.W.	11
key	5
key space	5
knapsack cryptosystem	98
knapsack problem	97
known plaintext attack	6
Kolmogorov's consistency condition	2
Kraft inequality	49
L^3-Algorithm	104, 107
Lagarias and Odlyzko attack	105
Lagrange interpolation	112
language	2
lattice	105
lcm	120, 142
least common multiple	120, 142
Legendre symbol	82, 129
length	22
LFSR	22
linear equivalence	29
linear feedback shift register	22
logarithm problem (discrete)	67
Markov chain	4
McEliece cryptosystem	94
McMillan inequality	48
minimal characteristic polynomial	30
minimal polynomial	154
Möbius function	132
Möbius inversion formula	133
modulo	124
monic	144
multiplexing	34
multiplicative function	126
multiplicative inverse	139
multiplicative Möbius Inversion Formula	134
mutual information	44
n-gram	2
NP complete	95, 97
NQR	82, 129
one-time pad	14
one-way function	64

operation	137
order	139, 141
O-symbol	67
out-of-phase autocorrelation	20
passive cryptanalysis	5
perfect secrecy	45
period	20, 26
periodic	20
plaintext	2
plaintext source	2
Playfair cipher	15
PN-sequence	24
Pohlig-Hellman Algorithm	69
polyalphabetic substitution	10
polynomial	142
- characteristic	24
- cyclotomic	155
- irreducible	26, 142
- minimal	154
- minimal characteristic	30
- primitive	26, 154
- reciprocal	24
- reducible	26, 142
prefix code	48
primality test	
- Cohen and Lenstra	84, 88
- Solovay and Strassen	84
prime	119
prime number generator	81
prime number theorem	81, 120
primitive element	26, 148
primitive polynomial	26, 154
primitive root of unity	148
principal ideal ring	143
Principle of Inclusion and Exclusion	134
privacy	1
product cipher	16
protocol	116
pseudo-noise sequence	24
pseudo-random	20
pseudo randomness tests	21
public key cryptosystem	63
QR	82, 129
quadratic residue	82, 129
quadratic non-residue	82, 129
quadratic reciprocity law	132
Rabin's variant to R.S.A.	89
randomized encryption	115
randomness postulates	20
reciprocal polynomial	24

reduced, y-reduced	106
reduced residue system	125
reducible polynomial	26, 142
redundancy	42
Reed-Solomon code	112
reflexivity	140
relation	140
residue class ring	141
residue system	
- complete	124
- reduced	125
ring	138
root of unity	148
rotor machines	16
R.S.A. system	77
run	20
Scherbius, A.	16
secure channel	5
Shamir attack	101
Shannon theory	39
shift register	22
shortest vector algorithm	107
signature	1, 65
simple substitution	8, 42
smooth	73
Solovay and Strassen primality test	84
source	2
source coding	47
spectral test	21
state	22
stationary	4
stream cipher	16
subfield	139
subgroup	138
subring	139
substitution	
- simple	8
- polyalphabetic	11
superincreasing	97
SV-algorithm	107
symmetry	140
tampering	5
text	2
threshold scheme	111
totient function (Euler's)	26
trace	158
transition matrix	4
transitivity	140
transposition cipher	15
trapdoor one-way function	64
U.D. code	47
unconditionally secure	45
unicity distance	42
uniquely decodable code	47
unit-element	137
Vernam cipher	14, 46
Vigenère cipher	9
Vigenère table	10
y-reduced	106
zero divisor	139
zero knowledge proofs	116